The Impact Clan

The Impact Clan

A Disenchanted Look at Asteroid Impact Monitoring Science

Germano D'Abramo

This book has been typeset with LaTeX

Science shines because it has found order in apparent chaos.
But chaos lurks behind apparent order.
Our worst hubris is taking for granted that Nature is intelligible.
Isn't this the last outpost of anthropocentrism?

Contents

Acknowledgements

I thank Alexander Unzicker (Pestalozzi Gymnasium, Munich) for useful advice and support. I owe the inspiration for part of the book to Valentina Borelli.

Chapter 1

Introduction

Objection to scientific knowledge: this world doesn't deserve to be known.

Emil Cioran

Wanting to use a scientifically appropriate jargon, this book is a critical, sometimes harsh, appraisal of asteroid impact monitoring science, which was born in the late 90's and in which I was involved from the beginning, although not in a leading role. This book, however, is also about my decade-and-a-half-long research experience in the field of astrophysics and planetology (although I have made research contributions also in other fields): hidden between the lines, the reader may find the personal and professional log of a trip which has taken probably the best years of my early adulthood and has now come to an end. I eventually decided to quit this job and to devote myself to teaching physics in high-school.

Two weeks after having communicated to my last boss my decision to quit, I happened to hear a telephone conversation between him and the project manager of the activities in which I had lately taken part. I was in the same room as the project manager. He was commenting positively on the choice of the new person (a woman) who was to replace me. He said that the choice was a fortunate one also because this would have helped to restore "gender parity" in the team. Nowadays it's important, he said. I did not need it, but after having heard this I was even more sure of having taken the right decision.

Mine is ironically the reverse path traveled by a lot of people, at least in my country (Italy): just after graduation it may happen that you teach for some years in high-school then, if you are lucky, you manage to get a permanent position in a research institute or university.

Several reasons led me to the decision to quit. First, in all those years I was unable to obtain a permanent research position: that's surely my fault, but the business is not perfectly fair, to say the least. In 2011, I took part in an open competition for a permanent research position at the National Institute for Space Astrophysics (INAF), where I had worked since 1998 with fixed-term contracts. During my interview, the only female member of the Examination Board, whom I had known since starting work with my former boss in the late 90's, commented on my publication list. Among my publications in international peer-reviewed journals I also had papers on thermodynamics and mathematical topics. She complained, worried as only a woman and a mother can be, that I should have published only on asteroids if I wanted to have a permanent position as a researcher (in astronomy and astrophysics, of course). I could not really believe my ears. I suddenly had the impression of having got lost in the maze of bureaucracy, without any possibility of salvation. How can a person who calls herself a scientist say such things? What would you think if you were urged to do only paintings and not sculptures if you want to become an artist? The sad side of the story is that she was right.

On a personal side, probably influenced by the first reason, I lost interest in that kind of scientific research. It did not take too much time to realize that this was not Science, at least in the sense I had been used to when I was in high-school, and studied physics: people like Galileo, Newton, Maxwell, Carnot, Clausius and Einstein, to name a few, made contributions that, when understood, had a great emotional impact on me. You knew immediately by yourself that what they did was smart and ingenious.

Based on my experience, I cannot say the same for (current) astronomy, astrophysics, and planetology research: the way things are run makes it almost always painstakingly boring and uninteresting, although those who are inside in leading roles continuously repeat to themselves and others (in a sort of self pep talk among nerds) that what they do is the coolest and most important thing ever.

By the way, among the most enthusiastic people I have ever met in this field are "astrometrists": astronomers who spend their daytime observing blinking images of portions of the night sky, taken with a telescope and recorded as digital images, in order to spot a few moving pixels (the telescopic image of an asteroid/comet). They usually react like kids in a candy store when they find one. I do not know whether it is somehow related to that disposition, but an astrometrist former colleague of mine (well known in our field for having discovered more than twenty comets while working at the Catalina Sky Survey in Arizona) now firmly believes himself to be the reincarnation of a seafarer who took part in one of James Cook's scientific expeditions in 1769 (to observe the transit of Venus). A few years ago, he unbosomed to me that before buying food at the supermarket, he checks it against poisoning agents with pendulum dowsing. I think it is dangerous to stay awake at night for too long.

Another typical trait of asteroid/comet discoverers (but also of theoreticians) is that they usually talk with deep emotional ardor and enthusiasm among themselves and (more worryingly) to ordinary (normal) people by using an incomprehensible language made of alphanumeric codes like 2004 MN$_4$,

2009 FD, 2014 LU$_{27}$ etc., indicating specific asteroids. And they do it as if you understand perfectly the meaning of each and, above all, as if you value this stuff as highly as they do. Once, when I was at the wedding party of a colleague, one of my former bosses, whom I had not seen for a while, could not find anything better than to open the conversation with: "Hi, Germano.... Did you notice the strange orbit of asteroid 2013 UP$_8$?". If 2013 UP$_8$ is for you nothing more than a strange, forgettable sign, for those people it is like the name of an ex-girlfriend, about which they know everything (come on guys, it is only a piece of rock after all!). In fact, this strange coding is the standard way in planetology of (provisionally) naming a newly-discovered asteroid. The rules are not so straightforward[1] but, for instance, the first four numbers give the year of discovery, while the remaining is related to the month.

I guess that many of you may be fascinated by night-sky watching or even be amateur astronomers yourselves, and strongly disagree with what I am saying about astronomy and astrophysics research. But believe me, what you do for your intimate intellectual pleasure to trigger your imagination and will to know (do you remember Kant's famous quote "Two things awe me most, the starry sky above me and the moral law within me"?) is not even close to what you find yourself doing in professional astronomy research. I derived my greatest intellectual pleasure almost exclusively from my solitary reading of books (scientific or not) and papers (scientific or not) and from my solitary thinking over what I had read and experienced[2]. Almost every one of us is looking for a

---

[1]http://en.wikipedia.org/wiki/Provisional_designation_in_astronomy

[2]I don't want to seem presumptuous, but I am in pretty good company, so to speak. In "Science and Civilization" (*Out of My Later Years*, 1950), Albert Einstein wrote: "I lived in solitude in the country and noticed how the monotony of a quiet life stimulates the creative mind. There are certain callings in our modern organization which entail such an iso-

deeper meaning in his/her own life. It was clear to me from the outset where to find it: neither in dominance nor in exerting control over other people, neither in human relations nor in money, but in the knowledge of Nature. Unfortunately, I discovered just as quickly that being engaged in institutionalized and professionalized scientific research was not exactly the same as pursuing my goal: the current world of scientific research is not infrequently a mean world, characterized by power and money games, by envy, malice and shabbiness. It throws you directly into the midst of human folly, no more nor less than any other ordinary and practical human activity. Some might say that this is the price to pay for becoming an adult. Nowadays, being engaged in research in astrophysics and planetology[3] means being a good technician, good at team work (read: you must do enthusiastically what someone else decides it's good to do). Furthermore, if you aim at higher positions, you need also to be good at public relations.

I think that what I am saying is not simply the blathering

---

lated life without making a great claim on bodily and intellectual effort. I think of such occupations as the service in lighthouses and lightships. Would it not be possible to fill such places with young people who wish to think out scientific problems, especially of a mathematical or philosophical nature? *Very few of such people have the opportunity during the most productive period of their lives to devote themselves undisturbed for any length of time to scientific problems. Even if a young person is lucky enough to obtain a scholarship for a short period he must endeavor to arrive as quickly as possible at definite conclusions. That cannot be of advantage in the pursuit of pure science.* The young scientist who carries on an ordinary practical profession which maintains him is in a much better position—assuming of course that this profession leaves him with sufficient spare time and energy. In this way perhaps a greater number of creative individuals could be given an opportunity for mental development than is possible at present." [emphasis mine]

[3]But I guess that it is more or less the same in other fields. I strongly recommend two recent and stunning books on the current state of astrophysics, fundamental physics and scientific research in general: López-Corredoira (2013) and Unzicker & Jones (2013).

of a romantic, nostalgic or elitist (or maybe naïve) person, but there are some epistemological truths here on how the current "Science System" is organized. If Science is what I think it is, how can you even imagine hiring a person from his/her first adulthood till retirement to "do" Science, as happens in almost all research institutes nowadays? I can understand hiring a person to teach at University and meanwhile, if he/she gets the inspiration after having long thought over the topics studied and taught, he/she can even arrive at small or great discoveries/advances in Science. They are primarily paid for their teaching service. "Being hired to do Science" is instead a cacophony to me, like a painter being hired to be, by contract, another Leonardo Da Vinci or another Michelangelo.

Science is a creative endeavor and you cannot be creative by contract. Furthermore, you cannot be creative by contract from the time of your engagement till your retirement, or at least not every employee in research can. Obviously, very few people are true scientists, just as very few people are true artists, and it follows that almost all the people hired in research institutes (who call themselves "scientists") are neither more nor less than average civil servants (if the research institute belongs to the Government) or R&D workers in a factory. Sooner or later, these people will stop being "creative" (if they have ever been) and they will end up spending their working hours doing something else, although still paid to be "scientists". Or worse, they keep doing the same stuff (or boring variations on the theme), publishing the very same papers over and over and thus contributing to creating and consolidating the opinion that the state of the art in Science is what they do.

Just belonging to the first category, several "scientists" working in my former research institute have spent almost their whole "career" chatting in the corridors, enjoying the thrill of being a "free mind" and devoting their working hours to their own hobbies like playing an instrument. Don't you know? If you don't play an instrument like Einstein, you cannot feel yourself a complete scientist! They are not scientists, they mimic scientists. Paid by taxpayers.

By the way, people love cliché because cliché is easy. And every period has its own cliché. I remember the media echo about one of the persons leading the team of engineers and physicists who "discovered" the Higgs boson: a physicist, a piano player and a woman (Fabiola Gianotti[4]). If she is a physicist and a piano player, then she is surely a big scientist. Because she is also a woman, she surely deserves the Nobel Prize. Luckily, people belonging to the Nobel Committee still have a grain of salt in their heads, and avoid (up to now) awarding prizes to mere technical managers involved in research (whether they play the piano or not). I do not want the reader to think that I am a misogynist. I want only to stress that, for me, critical thinking and adherence to truth outclass blind and collective adherence to cliché, gender issue fads, and so on.

In current scientific research a lot of time, for political and funding reasons, is spent on insignificant meetings (teleconferences) and giving presentations (a lot of presentations and less time devoted to research and inquiry mean presenting always the same stuff, with very tiny improvements each time). The most boring presentations ever are surely those intended for Education and Public Outreach (EPO): always the same, as a cliché, and usually built as a mega-advertisement for the group

---

[4]Read, for instance, this disgustingly apologetic newspaper article (in a newspaper that is second in Italy in terms of circulation), whose title reads more or less like this "Me, Between God and the Big Bang. Fabiola Gianotti, CERN Director-General, the Lady of the Universe": http://www.repubblica.it/scienze/2014/12/28/news/io_tra_dio_e_il_big_bang_fabiola_giannotti_direttrice_del_cern_la_signora_dell_universo_di_dario_cresto-dina-103841329/?ref=fb

8   GERMANO D'ABRAMO

you belong to[5]. In my humble opinion, this is not what public outreach should be. By the way, does public outreach make any sense at all? Who is it for? If you love Science (and if this is a true love, you are aware of it from your early childhood), you do not need ready-made public outreach sessions. Your curiosity already pushes you to critically deepen what you like most through books, papers and the Internet and eventually choosing specific courses at University. You surely do not need to attend public meetings where other people choose what is interesting for you. The only exception I am ready to make is EPO for children aged, let's say, between 5 and 12 years: good EPO sessions can trigger their curiosity and fascinate them. It must be said, however, that even at this age no truly curious mind needs anything flamboyant to be fascinated (e.g. the oft-quoted story of Einstein's compass).

In order to reach more people, public outreach is usually very low level. And if you don't care about Science at all, you will surely not change your mind because of EPO. So, the question is still who is EPO for? Probably, it is mainly for those people paid to do EPO. I have never met people more fanatical than those involved in astronomy research and outreach: they cannot even conceive that astronomy, or Science in general, might not be of interest. If it happens that you are not interested, you are a sick person to be healed. Western politicians want to export democracy, religious fundamentalists want to impose their "lifestyle", and current "scientists" want to diffuse their "Science". I believe there is a common drive beneath all these behaviors. Moreover, while making proselytes (the unconfessed goal of EPO) may make sense in religion, politics and business, it doesn't make sense at all in Science. Science is not democratic[6] and it doesn't need a ma-

---

[5]http://www.esa.int/spaceinvideos/Videos/2014/09/Notte_Europea_dei_Ricercatori_ESA-ESRIN_2014_-_Evento (from minutes 12:19 to 21:00).

[6]By the way, wanting to push by decree for the respect of minority

jority. However, let me now come to the topic of this book: impact monitoring science. What follows is a sketchy introduction; I will go into this more thoroughly in chapter 2.

Among all the asteroids in our planetary system, there are some with orbits that bring them close to the Earth. They are called Near-Earth Asteroids (NEAs) and belong to a more general class of objects (called Near-Earth Objects – NEOs) which also includes near-Earth comets. Obviously, if there are asteroids which may collide with the Earth, they are surely those belonging to the NEA population.

Impact monitoring and impact analysis of NEAs are carried out by automated collision monitoring systems that scan on a daily basis the most current asteroid catalog for possibilities of future impact with the Earth over the next 100 years (or in some special cases, even more). Whenever a potential impact is detected, it is analyzed and the results are immediately posted on some publicly available lists, called Risk Lists[7], where various information is given, such as the estimated size (and thus, the impact energy), the date of the potential impact and the probability of that impact. As the reader will see, this last information is the primary target of my criticism.

Impact monitoring activity is chiefly undertaken at the University of Pisa (where the monitoring system is called NEO-DyS-CLOMON2) and at NASA-JPL (National Aeronautics and Space Administration Jet Propulsion Laboratory), where the system is called SENTRY.

---

rights in Science (employment shares for ethnic, class, religious, linguistic or sexual minorities) appears a real idiocy to me. For instance, if you needed surgery for an important disease, you wouldn't want the surgeons appointment to have been facilitated by the fact that he/she belongs to one of the above minorities. You only want him/her to be extremely good at his/her job.

[7]NASA-JPL: http://neo.jpl.nasa.gov/risk/, University of Pisa: http://newton.dm.unipi.it/neodys/index.php?pc=4.1, ESA-ESRIN: http://neo.ssa.esa.int/web/guest/risk-page

Therefore, who does the title of this book refer to? Well, the Clan is the people of the two groups cited above. Some readers may be disturbed by this definition, especially because it refers to scientists, mainly mathematicians, working on a straight, but undoubtedly new, application of the old and honored discipline of Celestial Mechanics. As a matter of fact, an even more explicit term (Mafia) has already been used by one of the leading scientists belonging to the Clan in a talk during a conference session. As far as I remember, he humorously referred to the US research group as the "SENTRY Mafia" or "JPL Mafia" and to the Italian group as the "NEODyS Mafia" or "Pisa Mafia".

Leaving anecdotes aside, the term I used is appropriate in the sense that people belonging to it tend to act like a clan. If you are an informed layman or a scholar in the field, but not a particularly visible one, and you are critical towards what they do and you also have the possibility of publicly criticizing their research on forums and mailing lists, you are simply ignored. If instead you are a visible scholar (an astronomer, a planetary researcher) known in the relatively small community of planetologists, they usually publicly attack and even ridicule you as ignorant of the basic facts written in the long known scientific publications (theirs). See, for instance, what happened to Alain Maury[8] on the Minor Planet Mailing List (MPML)[9], where he boldly criticized their pro-activeness in announcing useless future "impact probabilities". He was not

---

[8] A French astronomer formerly working at the European Southern Observatory, now passionately involved in Education and Public Outreach in Chile. You can read the thread I am referring to at https://groups.yahoo.com/neo/groups/mpml/conversations/topics/29377. This was the public exchange. I know that the private ones have been even harsher.

[9] MPML is currently the primary mailing list for the minor planet (asteroid and comet) community (https://groups.yahoo.com/neo/groups/mpml/info).

too far from the truth, as I will illustrate in the rest of this book. Conversely, if you are an amateur or a professional astronomer supporting their work with follow-up observations[10] of NEAs included in their Risk Lists, they are fond of giving you and your work the greatest visibility (citing your contribution in their published papers, even striving to include your name and observatory in grant applications, usually from big agencies like NASA and ESA).

Then, if you are a young mathematician in your late twenties, possibly graduated under the guidance of one of the Bosses of the Clan, with fewer than ten publications (e.g., 9) most of which (e.g., 7) were in collaboration with the Boss, then you may have the chance to be appointed as a permanent researcher in the same University as the Boss; maybe by winning an open competition which other people applied for and who deserved the position at least as the winner, if not more, according to an unbiased reading of the minutes of the Examination Board (the secretary of which was obviously... the Boss).

Last but not least, every group striving to reach and maintain a leading role in something (arts, philosophy, medicine, science, knowledge in general, etc.) end up taking for granted that what they do is the state of the art in the field, without any possibility of criticism. They end up marching to their own drums. They may even become like some TV shows that are completely about other TV shows by the same broadcaster: something completely detached from reality.

Let me be clear from the beginning: this book will not be tender towards the whole business of impact monitoring, and I have done my very best to ground my point on objective and technical facts. Of course, I won't be so naïve as to claim

---

[10]Observations made in order to improve our knowledge of NEAs future orbital path. They obviously help to refine the "impact probability" calculation.

that my personal opinions, and sometimes some nasty ones, are not there. But I leave to the reader the not-so-hard task of spotting them. I hope that the more factual part of this book will achieve its goal of providing the reader with enough information to freely and fairly judge (and agree or not with) the more emotional part of it.

Chapter 2

The Near-Earth Asteroid population

There is nothing to understand, they are just stones.
A French astronomer
talking about asteroids.

In this chapter, I give a detailed (and a bit more technical, but don't be scared) overview of the population of asteroids we will deal with in this book. To follow this chapter, the reader is not required to have a PhD in celestial mechanics: almost all the content is accessible to everyone. Every now and then, when definitions or more technical concepts are introduced, I provide a few explanatory links in the footnotes.

The population of Near-Earth Asteroids (NEAs) is composed of those rocky bodies that can, according to their name, approach the Earth. They belong to a more general class of objects (called Near-Earth Objects – NEOs) which includes also near-Earth comets.

As many of you already know, comets are generally thought to be made of a mixture of ices (frozen gases including water vapor, methane, ammonia, carbon dioxide and hydrogen cyanide) and dust that was not incorporated into the planets when the Solar System was formed about 4.5 billion years ago. These objects have been traditionally classified into two groups, depending upon their orbital periods: Short Period Comets (SPCs), with orbital periods P around the Sun of less than 200 years, and Long Period Comets (LPCs), with P greater than 200 years. Among the SPCs, Jupiter family comets (JFCs) are further identified by their period $P < 20$ years, while the remaining comets, with $20 < P < 200$ years, are called Halley-type (HT) comets.

Unlike comets, NEAs are believed to be constituted by dynamically evolved fragments of main-belt asteroids (those confined between the orbits of Mars and Jupiter), entering the inner Solar System on chaotic orbits. The relatively strong gravitational influence exerted by Jupiter (plus some slow but unrelenting non-gravitational forces, like radiation pressure[1], the Yarkovsky effect[2], etc.) is the main cause of such a dynamical drift towards orbits that eventually lead these fragments to the inner Solar System.

I remind the most absent-minded reader that asteroids, like every body orbiting the Sun on a closed path, move around the Sun on elliptical orbits, i.e. ellipses with one focus on the Sun. An elliptical orbit is uniquely defined by six parameters called "orbital elements": semi-major axis a, eccentricity e, inclination i, longitude of the ascending node Ω, argument of periapsis ω, and epoch of the passage at perihelion t[3] (see Figure 2.1) The semi-major axis of an elliptical orbit is practically half its longest diameter or, the same as in celestial mechanics, the mean distance of the object from the Sun during one revolution. The reader should not worry here. He/She doesn't have to remember or even understand the meaning of every orbital element. He/She has only to know that these six

---

[1]http://en.wikipedia.org/wiki/Radiation_pressure

[2]http://en.wikipedia.org/wiki/Yarkovsky_effect

[3]http://en.wikipedia.org/wiki/Orbital_elements

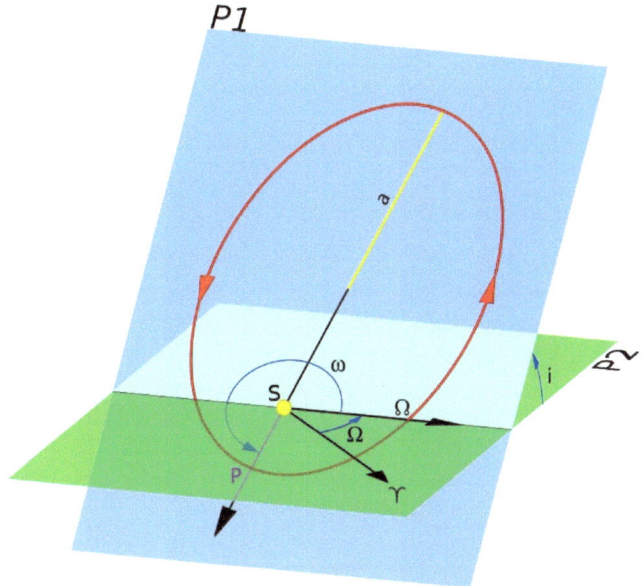

Figure 2.1: Schematic showing the orbital elements of a helio-centric orbit. P2 is the plane where the Earth orbits. P1 is the plane where the asteroid orbits. S is the Sun and P is the perihelion. (Source: Wikipedia Commons)

numbers uniquely identify an orbit, more or less in the way that three numbers (coordinates) uniquely identify a point (a place) in space.

Conventionally, the condition separating NEOs from generic comets and asteroids (e.g. those orbiting between Mars and Jupiter, and beyond) is given by perihelion distance q less than 1.3 AU[4]. The perihelion is the point in the orbit of a planetary object where it is nearest to the Sun, as opposed to the aphelion (Q), where it is farthest. From the orbital point of view,

---

[4]1 AU (Astronomical Unit) is the mean Earth-Sun distance (Earth semi-major axis), and it is nearly equal to 150 million kilometers, namely the distance traveled if you go around the world $\sim 3{,}700$ times.

near-Earth asteroids are usually classified into three main sub-classes (see Figure 2.2): Aten asteroids have a semi-major axis a less than that of the Earth and aphelia Q larger than the peri-helion distance of the Earth ($a < 1$ AU, $Q > 0.983$ AU); Apollo asteroids have a semi-major axis greater than that of the Earth and perihelia q inside the Earth aphelion distance ($a > 1$ AU, $q < 1.017$ AU); Amor asteroids are those with perihelia ap-proaching from outside the orbit of our planet ($a > 1$ AU, $1.017 < q < 1.3$ AU). Sometimes it may be found in the litera-ture that an NEA is classified as an IEO: IEO stands for Inner Earth Object and means an object that has aphelia smaller than the perihelion distance of the Earth. Its orbit is com-pletely inside that of the Earth.

Not all NEOs pose an immediate concern to the Earth. An NEO is conventionally classified as a Potentially Hazardous Object (PHO) when the minimum geometrical distance be-tween its orbit and the orbit of the Earth (the so-called Min-imum Orbital Intersection Distance or MOID[5]) is less than 0.05 AU (or about 20 times the distance between the Earth and the Moon) and it is brighter than absolute magnitude $H = 22$. For planetary objects, absolute magnitude H is a measure of the brightness an object would have if it were 1 AU from the Earth and the Sun and at the Sun-object-Earth angle of 0 deg. (if you think over that definition a bit you realize that this is an impossible configuration, but this is the formal definition). Brighter objects have smaller magnitudes. Absolute magni-tude provides also a measure of size, although not an intuitive one: greater absolute magnitudes mean smaller sizes[6].

---

[5]MOID is the minimum distance in three-dimensional space between the orbit tracks of the asteroid and the Earth. http://en.wikipedia.org/wiki/Minimum_orbit_intersection_distance

[6]The idealized diameter or size of an object (D, in km) is derived from the absolute magnitude (H) assuming an albedo (p) of 0.1 and using the following formula: D (in km) $= 1329\, p^{-1/2} 10^{-0.2H}$. The albedo p, or reflection coefficient, is the reflecting power of a surface. It is the ratio of

Figure 2.2: Near-Earth Asteroids, or NEAs, are those aster-oids with perihelion distance q less than 1.3 AU (1 AU is the mean Earth-Sun distance, and it is nearly equal to 150 million kilometers). Shown is a projection of NEA orbits onto the ecliptic plane (the Earth's orbit plane). That could give the wrong impression that actually all Apollo and Aten orbits in-tersect the orbit of the Earth. Obviously, this is not true since orbits are also inclined with respect to the ecliptic plane and rotated around the focus (the Sun). The outer bodies repre-sent asteroids in the main belt between Mars and Jupiter (not represented). (Source: Wikipedia Commons)

Objects in the PHO class are interesting since they might evolve into Earth impacting objects on a timescale of centuries due to the secular evolution of their orbital elements (change due to non-gravitational forces and gravitation). The aster-oidal component of the PHO subgroup is called Potentially Hazardous Asteroids, or PHAs.

---

reflected radiation from the surface to incident radiation upon it. As an example, an asteroid with $H = 22$ has an estimated size of ~ 170 meters, while an asteroid with $H = 18$ has an estimated size of $\sim 1,000$ meters.

NEOs have orbits that bring them very close to the Earth with a range of relative velocities between 10 and $30\,km/s$ for the asteroidal component (the typical velocity is about $20\,km/s$) and about twice these values for the cometary one. To have an idea of how large these velocities are, think of these values as dozens of times the speed of sound in the Earth's atmosphere at sea level.

Although in the case of impact comets can be many times more damaging than asteroids of similar sizes (higher relative speeds), it has been estimated by various authors[7] that comets contribute only a few percent to the total impact risk to the Earth. Basically, the region of the Solar System where the Earth resides is visited by comets with a frequency that is orders of magnitude less than that of NEAs. This is one of the reasons why impact monitoring systems currently devote their attention only to NEAs and not to comets. The other reason is that it is almost impossible to apply their monitoring algorithms to objects that suffer strongly from non-gravitational forces. Comets are active bodies. When they are relatively close to the Sun, their heated surface emits dust and gas and this modifies almost unpredictably their path in the long run. This further source of uncertainty cannot be easily dealt with in the current impact monitoring systems.

At the time I am completing this chapter (August, 2014), about 1495 PHAs are known (152 of which have diameters larger than or equal to one kilometer) out of a total of 11298 known NEAs (see Figure 2.3). These numbers are constantly growing since an average of 3 to 5 NEAs are discovered every day (even this average is bound to grow due to the fact that several research institutions and space agencies are investing money in ever more efficient discovery surveys).

The NEA population includes asteroids of all sizes: from

---

[7]The Spaceguard Survey: Report of the NASA International Near-Earth-Object Detection Workshop. Morrison, D., Editor, 1992, NASA.

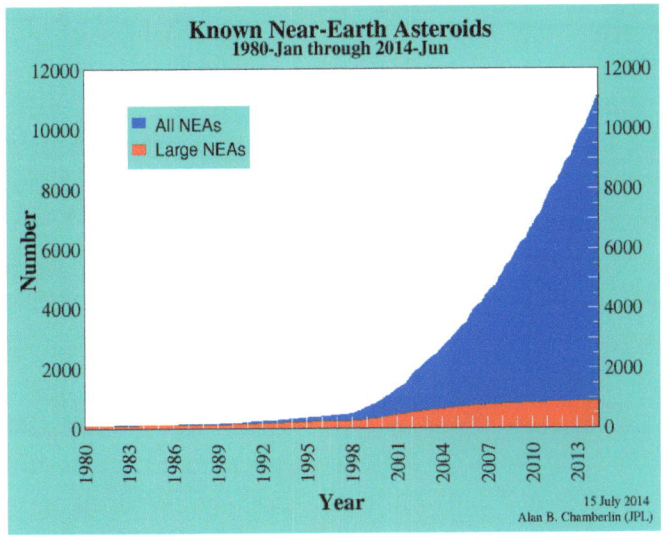

Figure 2.3: Known NEAs per year. "Large NEAs" in the above plot are kilometer-sized NEAs. (Credit: NASA/JPL)

tiny bodies of about 1 meter (and less) to few giant NEAs of diameter larger than $10\,km$. Some studies on the comparative risk analysis of impact hazard at the beginning of the 90's (e.g. Chapman and Morrison, 1994) have suggested that the greatest risk for mankind is associated with large impacts: a strong rationale was presented that NEAs with $D \geq 1\,km$ are the most dangerous and deserve the highest priority for detection (such risk estimates obviously being uncertain, the probable range is from 1 to $4\,km$ diameter). The diameter of $1\,km$ is taken conventionally as the catastrophic threshold, namely the threshold between enormous but regional effects and global scale effects in the event of impact. It is believed that if an asteroid more than $1\,km$ in diameter impacts the Earth, then probably more than 1/4 of the world's population would be killed. But no one can really say if these figures are meaningful or not.

The discovery and cataloging of at least 90% of all NEAs larger than $1\,km$ within 2008 was known as the "Spaceguard goal": this was the main result of a NASA study carried out after the 1992 U.S. Congress request to the NASA Administrator to submit a Program Plan to locate all NEOs greater than $1\,km$ in diameter. In the last few years both the U.S. and Europe have recommended the extension of the Spaceguard goal to smaller objects (equal to or greater than 140 meters in size) and called for the construction of a new class of big telescopes to accomplish the more demanding task.

But what does it mean that NEAs could pose a threat to the Earth? In what practical sense might they represent a danger in our lifetimes or in the entire lifetime of mankind? Is it possible to give an order of magnitude of the timescale of such events? As for earthquakes, what counts are the magnitude (energy) of the impact, directly related to the size of the impacting asteroid[8], and the statistical frequency of impacts for every size; and as for earthquakes, between these two quantities there exists a correlation.

Of course, if you know exactly the orbits of every asteroid, you are able to know with certainty (probability 1) whether there is any that will impact the Earth and when. But most of the NEA population is still unknown (especially for sizes less than one kilometer), thus our knowledge can only be statistical. Notice that statistical information always keeps its own value: suppose that you have a deterministic mathematical law which allows you to know exactly the position and the gray-scale level of every pixel in a picture. The analog of statistical information in this case is the overall view of the picture, which gives you a piece of information on a different level that you could never have with the deterministic law only.

For earthquakes, we will probably always rely on statistical

---

[8]The impact speed is also important, but its order of magnitude is more or less always the same.

information, since they appear to be definitely non-predictable. For asteroid impact, the knowledge of their orbits to a suitable degree of precision will be enough to say if they are safe or not for the next centuries (remember that asteroid orbits are slightly but inevitably and unpredictably modified by gravitational and non-gravitational forces on such a timescale).

All the statistical information about asteroid impacts is encompassed in what is called, in asteroid science jargon, *background impact probability (or frequency)*: this gives impact frequency vs. size. I think it is useful for what comes next to try to understand how it is derived. Here, I ask the reader to bite the bullet a bit: I need to introduce at least a few simple mathematical concepts, formulas and plots.

Astronomers first estimate the "per object" mean impact frequency. This is done by generating a synthetic population of point-like objects, which is thought to be representative of the overall orbital distribution of the actual NEA population around the Sun. Next, they numerically integrate over their motion and study the frequency of close encounters with the Earth. By extrapolating these statistics down to the Earth's radius and dividing by the number of point-like objects, they obtain the "per object" impact frequency (which, according to the most recent calculations, amounts to $\sim 1.6 \times 10^{-9} \, year^{-1}$).

The background impact probability is then obtained by multiplying the "per object" impact frequency by the estimated number of NEAs in different size ranges (the so-called "size distribution"). If a single asteroid has an average impact probability per year of $\sim 1.6 \times 10^{-9} \, year^{-1}$ and there are N of them around, what is the probability per year of being hit by any one of them? It is simply $\sim N \times 1.6 \times 10^{-9} \, year^{-1}$.

The NEA *size distribution* is obtained with several techniques which make use of observational data and extrapolation procedures beyond the scope of this book and it is characterized, mathematically speaking, by a power-law behavior: in simple terms, objects greater than a certain size are nearly five times more numerous than objects greater than twice that size. The background impact probability can then be approximated by the following power function:

$$\rho_i(\geq D) = 20 \times D^{-2.4}\, year^{-1},$$

where D is the diameter of the asteroid expressed in meters. Rigorously speaking, the background impact probability gives the average number of NEAs larger than a given size D (called the "cumulative" distribution) that hit the Earth per year. If this number is, say, less than 0.1, as happens with bigger asteroids, then it expresses a *mathematical probability* (per year). Otherwise, if it is greater than or equal to unity, as happens with (sub) meter-sized NEAs that fall on the Earth every year more than once, then it expresses the impact rate of objects greater than or equal to that size[9]

Accordingly, the average number of NEAs that hit the Earth per year with diameters between D and $D + \Delta D$ (e.g. between 10 and 20 meters) is given by the difference:

$$\rho_i(D; D + \Delta D) = 20 \times \left(D^{-2.4} - (D + \Delta D)^{-2.4}\right) year^{-1}.$$

The reciprocal of the background impact probability $\left(\frac{1}{\rho_i(\geq D)}\right)$ is the *impact interval time*, and it gives the "mean time" between two consecutive impacts of asteroids larger than or equal

---

[9]Wanting to be rigorous, the asteroid impact process is close to a Poisson process. The probability $P(N, D)$ that N asteroids of size greater than or equal to D hit the Earth per year is equal to $\frac{\rho_i(\geq D)^N}{N!}e^{-\rho_i(\geq D)}$, where $\rho_i(\geq D)$ is the background impact frequency. If $\rho_i(\geq D) \ll 1$, then the probability that *at least* one asteroid of size greater than or equal to D hits the Earth per year is $1 - P(0, D) \simeq \rho_i(\geq D)$ (it is enough to make the Maclaurin expansion of $P(0, D)$, the probability that no asteroid hits the Earth per year).

to a given size D. To figure out why this is so, think of a light bulb that is on, say, every half an hour, on average. This is the analog of the "mean time" between two consecutive events. Now, what is the frequency of this phenomenon (light bulb on)? It is obviously 2 *per hour* or $2\,hour^{-1}$, which is exactly the reciprocal of "half an hour".

As is clear from Figure 2.4, the constant power law is only an approximation of various data about NEA size distribution coming from different sources: the points plotted in Figure 2.4 represent un/published estimates of the NEA population in different size ranges. It must be said that every estimate of size distribution (and thus of background impact probability) has its own intrinsic uncertainty, which is often large and may not be fully characterized.

Besides formulas and plots, I wish the reader to remember the following simple facts (direct consequences of the power laws given above):

- Small asteroids fall on the Earth more often than big ones (basically because there are many more small asteroids than big ones);

- Asteroids less than ~ 5 meters in size are expected to fall on the Earth every year or more than once in a year, on average;

- Asteroids ~ 100 meters in size are expected to fall on the Earth every $\sim 3,000$ years, on average;

- Asteroids ~ 1 kilometer in size are expected to fall on the Earth every $\sim 800,000$ years, on average.

Please remember that, given the uncertainty at play, what really counts in the above numbers is the order of magnitude of each timescale, not its exact value. Remember further that "every X years, on average" should not be mistaken for "exactly every X years". It may happen that two 100 meter asteroids fall two times in a row, or even that they do not fall for 1 million years: these two possibilities, though, are less likely. Think what happens when you toss a coin. A head (or tail) is expected every two tosses, on average, but it may well happen that you come up with five heads in a row; it is only less probable.

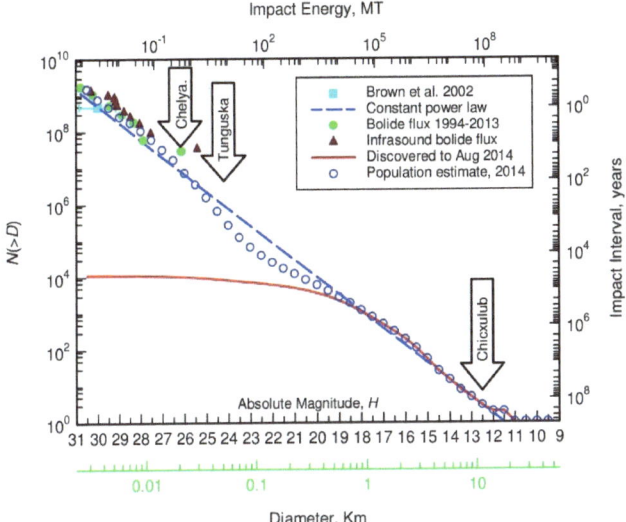

Figure 2.4: Blue-dashed line and blue-open dots: estimates of the cumulative population of near-Earth asteroids (NEAs) versus diameter (and absolute magnitude H, namely brightness at standard distance of 1 astronomical unit from Earth and Sun). $N(>D)$ is the cumulative number of objects with D greater than a given value. The fraction currently detected (red line) is nearly complete to $D \sim 2\,km$ (corresponding to absolute magnitude of about 16), but falls off rapidly with decreasing diameter, since smaller objects are harder to detect. There is also a scale (right) for the expected impact interval in years. Dots, triangles and squares represent some un/published estimates of the NEA population. The straight dashed line is a simple power law that approximates the estimates: with respect to the right-hand vertical scale, this line represents $\left(\frac{1}{\rho_i(\geq D)}\right)$, where $\rho_i(\geq D)$ is the first equation in the text. The estimated size and impact frequency of the asteroids that are thought to have caused Tunguska, Cretaceous-Tertiary (Chicxulub) and Chelyabinsk impact events are also indicated. (Courtesy Alan W. Harris)

Chapter 3

The Impact Monitoring Systems

Noise proves nothing. Often a hen who has merely laid an egg cackles as if she laid an asteroid.

Mark Twain

As I have shown in the previous chapter, in order to evaluate the risk coming from the still unknown NEA population we must rely on statistics and this piece of information is mathematically well represented by a power-law formula. I have also stressed that for the greatest part of the already known asteroids (whose orbits are very well known) we could already say with probability one (certainty) whether they will impact the Earth or not in the future. It is a matter of "relatively" simple celestial mechanics. None of them will do at present.

However, there is a subset of the already known NEAs (in the sense that it is composed of objects that have already been discovered and have got their provisional designation, like 2014 QN_{266}, remember?) whose orbits are still not very well known. Usually, they are those discovered recently and/or with a number of observations which are still not enough to calculate a "very good" orbit. It may happen that among them are some that will approach close to the Earth in the future, but due to the "non-perfect" knowledge of their orbits, we cannot exclude that these close approaches are indeed impacts. As a matter of principle, nobody can exclude that among them one lurks that is heading straight toward the Earth. Impact Monitoring Systems or Impact Monitoring Science deal with exactly these objects. In this chapter, vague concepts like "good" or "very well known" orbits will be clarified.

Let's start from the beginning, the discovery and tracking of an NEA. When a new NEA is discovered by telescopic surveys around the world, a preliminary orbit is computed using its available positions in the sky over a suitable (minimal) interval of time. This is the reason why my colleague astrometrists are so excited when they find a few moving pixels on digital blinking images. These luminous dots are the telescopic images of asteroids moving in space. The whole process, observation and measurement, is called "astrometric observation". More precisely, astrometric observations (or astrometric data) are the optical/radar measurements of the asteroid sky position with respect to the so-called "fixed stars" (the Earth, in the case of radar observations). These data are compiled from all the observatories around the world and made available at the Minor Planet Center (MPC)[1] at Harvard University. From these measurements it is possible to determine the path (also called "orbital solution") traveled by the asteroid in space (the first general orbit determination algorithm was developed by Gauss at the beginning of the 19th century).

Like every physical measurement, astrometric ones are affected by errors that make the resulting orbital solution un-

---

[1] The MPC designates minor bodies in the Solar System and has international responsibility for the efficient collection, computation, checking, and dissemination of astrometric observations and orbits for minor planets and comets (www.minorplanetcenter.org/iau/mpc.html).

certain to some variable degree. Sophisticated mathematical and numerical tools are now available to allow the propagation of the measurement errors to the six orbital elements $(a, e, i, \Omega, \omega, t)$. Let me recall from the previous chapter that an asteroid's orbit is mathematically identified by the six numbers $(a, e, i, \Omega, \omega, t)$ and can be visualized as a single point in a six-dimensional abstract space.

Unfortunately, due to error bars in every orbital element, the new NEA, soon after its discovery, is not represented by a single point in this six-dimensional abstract space; rather, it is represented by an uncertainty region, a six-dimensional volume with blurred contours. The reader should not try to figure it out mentally, as it is impossible; it may help to think of it as if it were an ordinary cloud in the ordinary three-dimensional world.

Consider the following simpler example. In physics, the length of a rod, when measured with a ruler of a certain resolution, is usually expressed with an error bar, e.g. $(1 \pm 0.5)\, m$. This means that, according to what we know, it can be anything between $0.5\, m$ and $1.5\, m$. Thus, the length, any length (actually, any measurement), is not a single value, but an interval. When you have three measures on three different axes (Cartesian coordinates), you have a three-dimensional interval, i.e. a three-dimensional volume. This line of reasoning can be easily extended to N dimensions, with N greater than 3.

When additional observations become available the volume of the uncertainty region changes. It usually shrinks since, in these cases, the error bars are reduced and the orbit estimate is refined. Measuring the asteroid positions in sky over a longer time span (i.e. more observations) allows us to better constrain its orbit (and shrink the uncertainty region). The extrapolation of a trajectory from a lot of positions all close to each other and all with their error bars is always more uncertain than an interpolation over the same (or a smaller) number of points, but spread over a longer arc of the trajectory.

When the nominal orbit (i.e. the orbital solution which mathematically best fits the observations) of the new NEA is geometrically close to the orbit of the Earth (small MOID – does the reader remember this acronym from previous chapters?) and the asteroid is expected to have one or more close approaches to our planet in the future, some orbital solutions that lead to a future collision cannot be excluded only on the basis of the available astrometric observations. This means that orbital solutions that lead to a collision are inside the uncertainty region and are fully compatible, within the errors, with the available astrometric observations of the discovered NEA. The "true" orbit of the discovered asteroid can be one of these.

As a loose analogy, if there is a rolling stone heading towards you at high speed and the determination of its trajectory comes with a big error bar that includes your position, in principle you cannot exclude that it will crush you (see picture 3.1).

In these cases, monitoring systems sample the uncertainty region with an appropriate number of sample points. They then evaluate the relative probability that the "true" orbit of the asteroid is one of the collision ones: this probability roughly corresponds to the ratio between the number of dots corresponding to colliding orbital solutions and the total number of dots with which you have decided to sample the entire six-dimensional uncertainty region.

Henceforth, we will refer to this probability with the symbol V_i. The collision orbits are nowadays commonly called Virtual Impactors (or VIs). Sometimes, V_i is also referred to as the VI impact probability. It is just this "impact probability" that will be the target of my criticism in the next chapter.

If a newly discovered NEA exhibits VIs, then it is promptly

Figure 3.1: The rolling stone analogy.

addcd by monitoring systems into their publicly available Risk Lists, together with its estimated probability V_i and other parameters like the Torino and Palermo Scale ratings that I will describe in a while.

Monitoring systems CLOMON2 and SENTRY apply their algorithmic procedures on a daily basis to astrometric data of all NEAs (both newly discovered and already known) observed the previous night by professional and amateur astronomers. They find tens of NEAs with VI orbital solutions every year, several per month on average.

Every time additional astrometric observations become available, the characterization of the asteroid orbit improves and the estimated impact probability V_i is re-computed. This may happen in the weeks, months, and even years following the discovery date. A typical pattern is that as the orbit becomes more precisely determined, the impact probability V_i

often increases initially, but then decreases until it falls to
zero, or some very low number. The reason for the initial
increasing behavior is not difficult to understand: since the
uncertainty region generally shrinks with new additional observations, some VI orbital solutions often remain inside the
uncertainty region in the orbital element space. Following the
stone example, if you manage to reduce the uncertainty with
which you know the trajectory of the rolling stone but your
position is still inside the error bar... well, the probability of
you being crushed has formally increased.

By the way, I ask the reader to think over the following
simple observation. If "impact probability" V_i fluctuates in
this manner, how can it be considered to be the actual impact probability? The way in which it fluctuates cannot be
predicted, otherwise impact monitoring scientists could know
from the beginning if it will go to 0 or to 1: the orbital information is the same for everyone and they cannot do magic.
As a matter of fact, *the true impact probability is instead the
probability that V_i goes to 1 as soon as we know the orbit of
the NEA under analysis with the necessary precision.* In some
sense, the true impact probability is a "probability of a probability".

Contrary to what may appear natural according to the
formal definition, if $V_i = 80\%$, it does not mean that there is
an 80% chance that V_i goes to 1 as soon as the orbit becomes
very well known. This is because of the underlying statistics. If
you know from statistics that the rolling stone crushes people
in your position once in 100,000 years on average, and the
uncertainty in your last updated trajectory measurement tells
you that there is an 80% chance of being hit (you fill 80% of
the entire error bar), what do you think is the *true probability*
of being hit? Is it 80% or the figure given by statistics (almost
0)? Do you trust more the way things usually go (statistics) or
your *still uncertain* measurement? Isn't a simple near miss the

more probable outcome? The uncertainty is in the information you gathered with your limited means upon a specific case, while statistics tells us how Nature usually behaves. This is an interesting mathematical-philosophical dilemma. It will be further investigated in the following chapters.

The V_i fluctuating behavior has been and is the reason for a lot of media hype. Things usually go like this: the Clan post a new possible impactor on their sites, then not unusually its "impact probability" increases with new astrometric observations coming in the following days and the Clan and the media react as if we are seriously in danger. Some time later, with additional astrometric data, the "probability" falls to an insignificantly tiny value and nobody remembers what they were talking about earlier. This is a game that reminds me of the overreaction of a bunch of chickens whenever the farmer makes the same old innocuous noise.

Consider, for instance, what happened with asteroid Apophis. I guess that almost every one of you knows who this guy is, but let's tell its story anyway. (99942) Apophis is a piece of rock ~ 300 meters wide discovered in June 2004 by astronomers Roy A. Tucker, David J. Tholen and Fabrizio Bernardi. Its provisional designation was 2004 MN_4 and it was then named after an ancient Egyptian evil demon (great sense of humor!).

On 23rd December 2004, monitoring systems calculated an initial impact chance of 1 in 233 for the year 2029. Later that day, based on a total of 64 observations, the estimates were changed to 1 in 62. With new observations coming in over the very next days, on 27th December 2004 the reported chance of impact reached its highest value of $\sim 1/37$. A really worrying one. Then, after even more observational data were gathered, the impact in 2029 was ruled out. Currently, Apophis has a chance nearly equal to $\sim 5 \times 10^{-6}$ of hitting the Earth in 2068. Statistics tells us that objects greater than or equal to

~ 300 meters fall every 40,000 years, on average.

Now, let me make a not-so-short digression. If you want to easily evaluate and communicate an impact hazard, you need a Scale. You cannot provide cumbersome probability values or energy yields which are usually expressed with powers of ten (try for yourself – what does a probability of 4.49×10^{-8} and an energy yield of 5.53×10^7 joules mean for you?). The asteroid impact scientists came out with two risk scales: the Torino and Palermo Scales. These are named after the two Italian cities where they were first proposed during as many meetings.

The Torino Scale (TS) was created by Richard P. Binzel (MIT). It was officially adopted at the Impact Workshop held in Turin in 1999. It was a revised version of a scale already presented by Binzel at a United Nations conference four years earlier. I participated in the Turin meeting and it was one of the first times for me to be among such highly qualified and selected people. Actually, the meeting did not always go smoothly. I remember people shouting each other down during some sessions. I remember one member of the Clan complaining animatedly to the late Brian G. Marsden (the former director of MPC) who, according to him, was unnecessarily delaying one of his first publications on asteroid impact monitoring, of which Marsden was nominated as one of the referees.

In order to classify the hazard posed by an NEA with non-zero "impact probability" V_i, Binzel created a two-dimensional diagram with the "probability of impact" on the x-axis and the "estimated impact energy" (depending on the size of the NEA, as already said) on the y-axis. Then, he commonsensically divided the plot area into eleven sectors (from 0 to 10), which he colored with a color code made of bright colors from white to red (see Figure 3.2). This way, every NEA with non-zero V_i necessarily falls into one sector, whose number is an easily

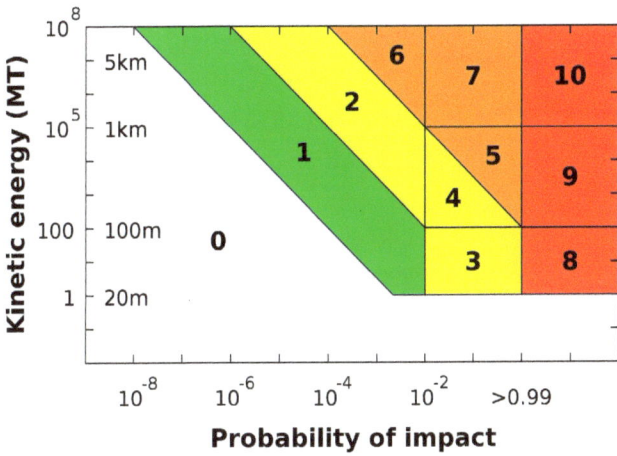

Figure 3.2: Torino Scale. The scale in meters (kilometers) is the approximate diameter of an asteroid with a typical collision velocity. (Source: Wikipedia Commons)

understandable indicator of hazard severity. The Torino Scale can be regarded as the "Richter scale" of asteroid impact hazard.

This scale has several drawbacks. Some technical flaws were spotted, not long after its presentation, by Spanish mathematician Joaquín Pérez, working then at the Economics Department of the University of Alcalá de Henares. The more intuitive one is that a very likely (80%–95%) collision with an energy of a hundred million megatons is rated as less threatening than a certain one megaton collision. Moreover, since size and impact energy are almost always known with large uncertainties, if for instance an object happens to fall close to the boundary point between sectors (going clockwise) 4, 5, 9, 8 and 3, then we really do not know if it is a moderately worrying case (3) or a really frightening one (9).

However, the real blow to this scale has been given by

passing time. This scale was chiefly thought of as a communication tool for the general public, but although the man in the
street knows very well what a 6th grade earthquake means,
practically no one knows what Torino Scale 0 or 2 means.
Earthquakes happen every year more than once and almost
everywhere. They cause casualties. No one in at least 4,000
years has ever experienced as devastating an asteroid impact
(leaving aside minor meteor impacts). What do you do with a
scale that almost always gives you grade 0? Moreover, while a
5th grade earthquake is something that has really happened,
a TS 4 asteroid may become within two days of astronomical
observations a TS 0. Actually, this is what happens almost
all the time. Nowadays, the most zealous users of TS are obviously its creators, the people belonging to the Clan. Again,
they march to their own drums.

A few years later, in 2001, a new hazard scale was proposed: the Palermo Technical Impact Hazard Scale, or simply Palermo Scale (PS). I guess that I do not have to say
where it was presented. I participated also in this meeting.
This time the mood was more relaxed, probably thanks to the
great food, wine and weather of Southern Italy. It's a hard
life for researchers, isn't it? The Palermo Scale is not a graph-
ical scale and it was not thought of as a communication tool
for the general public. It is used only among asteroid impact
specialists. It does not give a discrete rating, but a continuous one and with no color codes. Let me sketch here how
a PS rating is calculated. Suppose an NEA with size D and
non-zero "impact probability" V_i is found. They calculate the
ratio between V_i and the background impact probability that
an unknown object of the same size has of hitting the Earth
between now and the impact date of the known NEA under
analysis (time span ΔT). I hope the reader can easily see that
this last number can be obtained by the product of the background impact probability $\rho_i(D; D + \Delta D)$ with the time span

ΔT. The Palermo Scale rating is then taken as the logarithm in base ten of that ratio:

$$PS(NEA) = \log_{10}\left(\frac{V_i}{\rho_i(D; D + \Delta D) \times \Delta T}\right).$$

The reason why they decided to use this logarithm is that this mathematical function gives roughly the exponent of its argument, if you write it as a power of ten: for example, $\log_{10}(1) = 0$ because $10^0 = 1$, $\log_{10}(100) = 2$ because $10^2 = 100$ and so on. The ratio calculated above can always be written in the power-of-ten notation, like 4.49×10^{-8}, and it is undoubtedly easier to communicate only the exponent (order of magnitude) rather than the whole number (be careful, as $\log_{10}(4.49 \times 10^{-8})$ is not exactly equal to -8 because you also have the factor 4.49, but it is close).

As their creators and most zealous users say, Palermo Scale 0.0 means the predicted event is essentially "as expected" in that time interval; positive values mean the predicted event is "extraordinary" and deserves attention. Negative values indicate the predicted event is only a small addition to the expected impact flux.

I strongly disagree with the wording *"positive values mean the predicted event is "extraordinary" and deserves attention"*. This is blatantly not true. As I will show in the next chapter, the true impact probability of an NEA included in the Risk Lists is almost always less than the background impact probability of objects of the same size.

Summing up, the only meaningful values of impact probability are: 0 for known safe asteroids, the *background impact probability* law for unknown asteroids and asteroids with uncertain orbital solutions, and 1 for known colliding asteroids (none at present).

Chapter 4

Their "impact probability" is not what they say it is!

The life of the enemy. Whoever lives for the sake of combating an enemy has an interest in the enemy's staying alive.

Friedrich Nietzsche

In the last two chapters, I hope to have provided the reader with all the basic information needed to understand the asteroid impact science and the impact monitoring business. Now the time has come to focus on my main criticism of all that. My thesis is simple: *the impact probability V_i calculated for the subset of NEAs with "poor quality" orbits is not the true impact probability of these objects.* It can be shown, and I will do so in this chapter, that the impact probability for these NEAs is not much different from the background impact probability; in fact, it is usually less than that. Either you can say with probability one that an asteroid will hit the Earth in the future, like with those with very good orbits, or you always have to rely on the background impact probability. It does not matter if the asteroid has been found to have Virtual Im-

pactor orbital solutions with high or low V_i. When you hear
that impact monitoring systems have spotted "one of the most
dangerous asteroids ever found[1]" (highest Torino and Palermo
Scale ratings ever), please don't bother. It's all nonsense.

Before getting too technical (relax, I'm exaggerating), I
want to invite the reader to think over the following two arguments; I think they may help understand without formulas
what is wrong with the impact monitoring systems logorrhea
about threatening objects.

First. Before 1999 the Earth was a relatively safe place to
live in, at least from an astronomical point of view; we did not
seem to be under attack by nasty asteroids. OK, we already
knew that every now and then (speaking in terms of geological
timescales) the Earth was being hit by some very swift "space
rock". However, before 1999 it was as if we were in a sort of
"space truce". Since 1999, the birth date of current impact
monitoring science, we have been under asteroid impact siege
almost every day. Almost every week a new NEA is discovered
which has non-zero "impact probability" V_i, and it is added
to the continuously growing Risk Lists. Periodically, one is
even discovered that is "one of the most dangerous asteroids
ever found". My deduction is: someone, after 1999, must have
broken the truce! Jokes aside, commonsense already tells us
what someone (the Clan) fails to acknowledge: the birth and
the continuous updating of the Risk Lists do not add an iota
to the actual risk coming from asteroid impact, which is and
has always been the same from the time man first walked the
Earth. It is just much ado about nothing.

---

[1]Google these words and see how often this phrasing has been used by
the media (newspapers, TV, Internet) for NEAs included in the impact
monitoring Risk Lists. I have countless times even heard members of the
Clan saying: "Look, this object is one of the most dangerous on the Risk
List. Astronomers must re-observe it as soon as possible and we need to
contact some of them urgently".

Second. What is going on with impact monitoring is more or less as follows: a zealous physician comes up with a new test for a very, very rare (say, a 1 in 7 billion chance) and extremely and excruciatingly fatal disease (as far as we know, this may very well be comparable to the frequency of human beings actually killed by an asteroid so far). If the test is positive, it means that you have a *non-zero* probability of being infected: it does not mean that you will *certainly* be infected, but if you are infected you will certainly have a positive response. Suppose now that the zealous physician struggles to apply his test to the Earth's whole population: he feels himself called to the mission of saving the world from this execrable disease (you already guess that there is something wrong with this guy...). Suppose, further, that 100,000 people (out of roughly 7 billion) get a positive result. Obviously, it can easily happen that none of them fall sick (all false positives), since the disease is very, very rare. However, now the "new" information given by the test contributes to unnecessarily and dangerously increasing the level of alertness in everyone, from ordinary people to decision makers. This may generate an unnecessary and dangerous hysteria about the disease: it may seem that it has become less rare (a lot of "positive" tests!), that there is an outbreak going on, and it can lead politicians and decision makers to take the wrong decisions. What really counts is the statistical incidence of the disease (already known by other means), which is always the same, despite the zealous physician's test results. This is more or less what could happen with asteroid impact monitoring science. Fortunately, there has been no hysteria so far: the only result obtained is bringing more visibility (and even funding) to impact monitoring science.

The Impact Clan seem to operate under a paranoid delusion. Every time their algorithms find a new asteroid with non-zero "impact probability", they act like the zealous supporters of McCarthyism at the time of the Cold War finding an impending (and non-existent) communist threat. They first communicate with each other in a highly confidential manner (via emails, phone calls). Then, after having reached agreement on the numerical values, they publish the new result on their web pages (quite often this information becomes news in the press – see the following chapter). Usually, they take such a decision with the same tension and urgency as if they were deciding the fate of the world (see footnote 1). It has happened the same way every time since the late 90's, although they couldn't not know, even without my criticism of the meaning of their probability, that all the objects in their Risk Lists are almost certainly harmless[2]. They seem to live under siege, in a world of their own, of their own making.

I would be curious in this case to ask the Clan to join Kant's bet, in order to measure how much they really believe in what they appear to be worried by. Philosopher Immanuel Kant wrote in his most famous work, *The Critique of Pure Reason*:

> "The usual touchstone, whether that which someone asserts is merely his persuasion — or at least his subjective conviction, that is, his firm belief — is *betting*. It often happens that someone propounds his views with such positive and uncompromising assurance that he seems to have entirely set aside all thought of possible error. A bet disconcerts him. Sometimes it turns out that he has a conviction which can be estimated at a value of one ducat, but not of ten. For he is very willing

---

[2]I have already shown that objects smaller than 5 meters have a high statistical probability of hitting the Earth once or twice per year, but their impact does not usually have a significant effect on the ground. For greater objects, the tiny statistical probability is a guarantee of their safety.

to venture one ducat, but when it is a question of ten he becomes aware, as he had not previously been, that it may very well be that he is in error. If, in a given case, we represent ourselves as staking the happiness of our whole life, the triumphant tone of our judgment is greatly abated; we become extremely diffident, and discover for the first time that our belief does not reach so far. Thus pragmatic belief always exists in some specific degree, which, according to differences in the interests at stake, may be large or may be small."

Just before quitting my involvement in impact monitoring activities, I had been appointed as Front Desk Operator at the ESA SSA NEO Coordination Centre[3] (NEOCC) in Frascati (near Rome). NEOCC aims to coordinate and to contribute to the observation of Near-Earth Objects in order to evaluate their impact hazard. NEOCC hosts advanced systems for orbit computation and impact monitoring as well as the tools and the data needed for performing risk assessment (e.g. NEOCC hosts the Risk List from NEODyS).

One of the reasons[4] why my presence there became unsustainable was that the people leading this center wanted me there 8 hours per day, for five days a week, while I asked for permission to have some spare (and unpaid) days for teaching. Their plan for the future is to become really operational and to have a lot of people to guarantee a 24/7 service. They want to be ready to cope with "emergencies", they said. They want at least one person able to answer the phone or reply to emails 24/7, complying with the needs of the service's customers (the general public, scientific bodies, international organizations and decision-makers)[5].

---

[3]http://neo.ssa.esa.int

[4]The main one is what is written in the whole book.

[5]Among the variegated tasks of the Front Desk Operator (FDO) was

I really do not understand where the "emergency" is. They are not a fire service operations center, they are not the 911 operations center, they are not the Civil Protection operations center. Can the reader imagine the following phone call? A housewife: "Hurry up, there is an asteroid falling in my backyard!!", NEOCC: "Please, don't panic, we'll come in a few minutes...". As a matter of fact, in the foreseeable future NEOCC will probably have only a function as an Education and Public Outreach center on asteroids, comets and fireballs, and a mild coordination role among professional astronomers in order to optimize astronomical observations.

Now, I want to go into a bit more detail, in order to give a quantitative proof of my main thesis. I have already done this in a paper published in *Chance* (peer-reviewed magazine of the American Statistical Association) in 2013. That article has had a troubled story. I will tell of it in chapter 7. Now, I ask the reader to follow me in the following thought experiment.

Consider NEAs in a particular size range $(D; D+\Delta D)$, such as $(30\,m; 40\,m)$. Consider further a suitably long time span in the future ΔT, say 1,000 years. According to what I showed in chapter 2, during that period of time we should expect the impact of $\rho_i(D; D+\Delta D) \times \Delta T$ objects with diameter between D and $D+\Delta D$, on average. By plugging in the previous numbers, we have that $\rho_i(30\,m; 40\,m) \simeq 0.003$, thus the expected number of impacts in the next 1,000 years of NEAs between 30 and 40 meters is 3, on average. As I said before, it can be more or less than that number, as this is just statistics. But let me stick to this number for the sake of the argument. Dur-

---

the checking and correction of official forums and mailing lists on asteroids against "bogus" statements: I see this as a real waste of time for a highly qualified person as the FDO is required to be. I really could not think of me chasing crackpots over the Internet, when everybody knows that nowadays any idiot can post what he/she wants in cyberspace. By the way, aren't the same Risk Lists a bunch of bogus statements?

ing the same period of time ΔT, impact monitoring systems (long live impact monitoring systems!) will surely find a lot of newly discovered NEAs with non-zero "impact probability" V_i in the same size range, including obviously the three that will actually hit the Earth. Can you guess how many in total? I have already made the computation going through the Risk List archives. I have found that till now monitoring systems find an average of ~ 8 NEAs between 30 and 40 meters per year. So, in 1,000 years, it amounts to $\sim 8,000$ objects.

Now, let's come to the central question: what is the probability that one of the $\sim 8,000$ asteroids indicated by monitoring systems as "dangerous" is actually one of the 3 asteroids that are expected to hit the Earth in the next 1,000 years? The answer to this question gives also the *true impact probability* of the objects in the impact monitoring Risk Lists. No matter how high or low is their V_i, the true impact probability of NEAs between 30 and 40 meters included in the Risk Lists is anyway given by the ratio:

$$W(30\,m; 40\,m) \simeq \frac{3}{8,000} \simeq 0.0004.$$

Besides, this value is independent of the chosen time span ΔT, because both the numerator and the denominator of that ratio have been obtained by multiplying ΔT by other stuff (the background impact probability and the annual frequency with which impact monitoring systems find "dangerous" NEAs, respectively).

It must be said that W *actually gives an upper bound to the true impact probability*. The true impact probability (in every size range) is less than or equal to W: the denominator of W, namely the number of asteroids indicated by monitoring systems as dangerous, is a fraction (almost constant) of all the asteroids discovered in the time span ΔT and the number of discoveries in that period is surely less than the total number of asteroids that pass close to the Earth and can be discovered. Thus, the denominator of W is surely underestimated and W is consequently overestimated. In fact, many unknown asteroids that pass close to the Earth remain too dim to be detected by telescopes. They are too small in size and/or still "too distant". Moreover, some NEAs are not found because telescopic observations miss them, as surveys do not image a portion of the night sky when they are there and bright enough to be seen. This becomes especially true for asteroids relatively close to the Sun, where observations are more sporadic or even impossible.

The calculation of the above ratio can be extended to all sizes to provide the upper bound W of the true impact probability for all the asteroids included in the Risk Lists. I did the full job in my *Chance* paper and the result is shown in Figure 4.1. Now, consider an NEA included in the Risk Lists. These lists usually report the impact year, e.g. 20XY, the probability V_i and the Torino and Palermo Scale values. It is interesting to compare W for that object with the background impact probability calculated for objects of the same size and for the single calendar year 20XY. To do that, we need to use the second formula of the background impact probability, that for size ranges (remember?):

$$\rho_i(D; D + \Delta D) = 20 \times \left(D^{-2.4} - (D + \Delta D)^{-2.4} \right) \, year^{-1},$$

since I computed W values in every single size range. Moreover, you have to formally multiply $\rho_i(D; D + \Delta D)$ by 1 (year), since we want the background impact probability for a single calendar year (20XY). If one considers the figures at their face value (Figure 4.1, the histograms give W values, the line gives the background impact probability values), you may see that for almost all size ranges, W is even less than the background impact probability: this means that no NEA in the Risk Lists is "dangerous" or "the most dangerous ever found". Rather, it is always less dangerous than all the still unknown NEAs in the same size range.

Ironically, what the Clan know to be most dangerous is actually less dangerous than what the Clan (and we) do not know at all. Not to mention that comets are not even tackled by the monitoring systems, as I said in chapter 2. OK, the hazard coming from them is thought to be "negligible" with respect to asteroids, but it is not zero.

There is a further irony in all this (or, maybe not): while people completely engaged in human affairs, like the Clan, who wallow around in the things of this world and seem to love human beings (they are surely not reclusive and misanthropic) continuously scare them, I, who sometimes wishes this despicable world were obliterated by a huge asteroid, am trying to reassure people that their announcements are hot air.

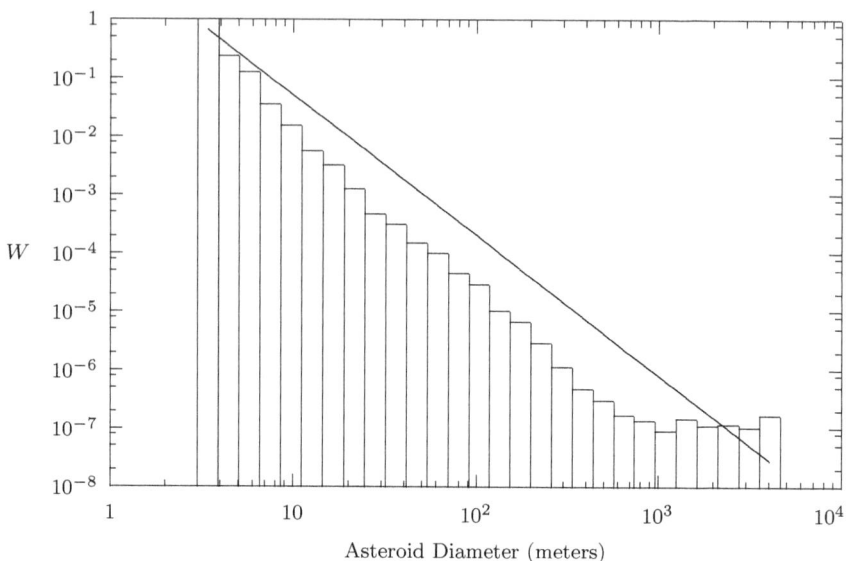

Figure 4.1: The upper bound W of the true impact probability of asteroids included in the Risk Lists, given in size ranges. The scales of the plot are logarithmic. The histograms give W values; the line gives the background impact probability values (calculated for a single calendar year) in the same size range. Note the first column to the left: this does not mean that asteroids included in the Risk Lists with diameter between 3 and 4 meters will certainly impact the Earth (probability 1). I have already explained in this chapter that W actually gives an upper bound to the true impact probability of objects in the Risk Lists. The true impact probability is probably even less than the values indicated by the columns.

Chapter 5

In the headlines: a love-hate relationship

Newsreader: A huge asteroid could destroy Earth! And by coincidence, that's the subject of tonight's miniseries.

Dogbert: In science, researchers proved that this simple device can keep idiots off your television screen. [TV remote control] Click.

From a Scott Adams comic strip

I am quite naïve, I admit. I have always thought that only big discoveries should get press coverage. OK, you will say, but what is a big discovery? How do you define a discovery as "a big one"? Yes, I get the point, but I think it is more or less like the definition of "intelligence". It is difficult to delineate, but easy to spot. Past examples may help. The discovery of "Medicean Satellites" by Galileo made it clear and difficult to refute that the world was actually not as previous scholars had thought for centuries. This was a big discovery. It was a paradigm shift, actually. And this gained, in some sense, the headlines of that time. The fact that Einstein was able to explain the perihelion precession of Mercury and successfully predicted light bending by the Sun is a big discovery. And it actually did hit the headlines of his time. By giving a look at the future, the unambiguous discovery of extraterrestrial life would surely deserve the headlines.

Nowadays, unfortunately, almost every scientist has made a press release on his own work at least once in his lifetime. Professional scientists are paid to do research (taxpayers, remember?), thus they have to re-sell what they have done, no matter if it is a paradigm shift or empty talk. And the more money they have received, the more flamboyant the press release should be. "Big Science", for instance, even has a press office of dozens of people (these people are also paid by taxpayers...). Consider what happened with the "Higgs boson". A lot of time and money was spent collecting a huge amount of data, and analyzing them with highly sophisticated (and not so crystal clear) statistical tools: the "Higgs boson" must be there. How can it be otherwise? Tears of joy and lots of champagne[1].

Impact monitoring scientists certainly do not think outside the box. As far as I know, the Clan do not have a dedicated press office yet, but more than once (a lot more than once, actually) their results have been in the headlines. It is enough for the reader to perform a simple search of the Internet. See, for instance, two articles posted on TheGuardian.com in 2005[2] and 2013[3], regarding that worn-out hobbyhorse of the Clan, asteroid Apophis. I urge you to notice the tone used in the phrasing. If you take care to re-read newspaper articles of this kind even just a few weeks after their publication, let alone years, you cannot fail to perceive their futility. My feeling is more or less the same I have when I happen to read old news on the last "viral" fashion spread over the Internet: what remains of it now? Almost always nothing. Trust me, try for

---

[1]http://www.youtube.com/watch?v=0CugLD9HF94

[2]http://www.theguardian.com/science/2005/dec/07/spaceexploration.research

[3]http://www.theguardian.com/science/across-the-universe/2013/jan/07/apophis-potentially-hazardous-asteroid-earth-wednesday

yourself. Seen from outside, all this has the consistency of a cyclical fad phenomenon and people working in this field are perfectly comfortable with that. Is this Science? Or, honestly speaking, is it more akin to the behavior of some businessmen who advertise and push for their business?

As a matter of fact, as I wrote in the title of this chapter, the actual relationship between impact monitoring people and the press is a "love-hate" one. More than once I have heard people from the Clan complaining about the way in which journalists deal with the technicalities of their work: "...they do not understand probability at all...", "...they are ignorant and oversimplify things...", "...they scare people too much...[*sic*]". Nevertheless, these complaints melt like snow in the sun whenever some journalist wants to talk to one of them or even conduct a TV interview about the last impact scare. They appear to act like atheists, excited and trembling before the Pope. Like almost every man, they are extremely vain. Obviously, they would take these words as extremely offensive and would reply that they are compelled to talk to journalists because what they do is extremely important and people need to know and journalists need to be educated.

So, where is the truth? Are journalists making such a fuss every time and are scientists always misunderstood, or are their words made more shocking for the sake of journalism? Or do asteroid impact scientists have at least some responsibility for this misunderstanding? Obviously, I have my opinion. When you meet a person for the first time and work with him (e.g. a scientist with a journalist), this person may disappoint you. It can even happen twice. But not the third time. If it happens again, either you are dumb or what he does looks good to you.

Maybe some of you remember what happened on the 15th February 2013. Early in the morning of that day (Central

European Time), a meteor ~ 17 meters[4] in size exploded over the Russian city of Chelyabinsk. I remember I was in my (now former) office in the ESA NEO Coordination Centre in Frascati and I saw the footage of the fall and explosion, that the reader has surely seen dozens of times. You cannot even imagine what happened in the coming weeks and months among asteroid impact scientists. This unexpected event (both the impact and the fact that it was recorded by dozens of dashboard cameras) was a tremendous boost for their activities. It was literally a stroke of luck. At least, this is how the Clan used to define it. Obviously, only during private conversations and instantly specifying that it was good that no one was killed; but then adding that it would be good for asteroid research if minor impacts like this happened more frequently. It would ensure more attention from the public. However, during public meetings following the impact, they showed up as extremely worried about the consequences it caused (one thousand people slightly injured by broken windows), and in a professional and serious tone explained that their work is aimed exactly at the prevention of such disasters. Is this a bipolar/schizophrenic behavior or simply a dreary one?

The irony in this case was that they did not foretell the impact at all, but that was not their fault. The Chelyabinsk meteor was simply unknown before it eventually entered the atmosphere. It belonged to the still unknown NEA population I described before. Thus, it was simply not possible for the impact monitoring systems to calculate its "impact probability". Had it been the case (as happened with the tiny asteroid[5] 2008 TC3), they would have surely spotted the oncoming impact.

---

[4]Objects of that size (15 to 20 meters) fall on the Earth once every ~ 60 years on average, although the probability that they will fall on a city is very much less than that.

[5]http://en.wikipedia.org/wiki/2008_TC3

On the very same day, the astronomical community and the media were busy with something else; they were actually "looking" in the opposite direction with respect to the Chelyabinsk meteor. It was long known by orbit computers that asteroid (367945) Duende (former provisional designation, 2012 DA_{14}) would approach very close to the Earth on the 15th February 2013. The asteroid, $\sim 45\,m$ in size, passed 27,743 kilometers above the Earth's surface, closer than satellites in geosynchronous orbit (well within the orbit of the Moon, obviously). Before its close approach, asteroid Duende was included in the Risk Lists with an "impact probability" of $\sim 5 \times 10^{-5}$ for calendar year 2026. After its close approach, the new astrometric data collected by astronomers showed that it was eventually a safe asteroid. If you search on the Internet for 2012 DA_{14}, you may see how extended the media coverage of this event was.

Let me express here a personal opinion on close approaches. They are quite common phenomena, above all when the approaching asteroid is a small one (smaller asteroids are more numerous, remember?). I really do not understand the worldwide fascination with such events in the media or the attraction prompted in people by the media on these things. I understand that when comets come close to the Sun and become visible from the Earth, it could be a nice show. But nothing more than that. Then think if you are not even able to see the approaching object with the naked eye, as almost always happens with asteroids.

Asteroid close approaches may be professionally interesting for astronomers, but they are certainly not exceptional events. By using the background impact probability rescaled for a bigger Earth (new Earth radius equal to 6,371 km+27,743 km), it can be calculated that close approaches to the Earth within 27,743 km of objects greater than or equal to $\sim 45\,m$ happen every ~ 16 years, on average. Can events of this frequency be considered "rare" or "exceptional"? By the way, just as I'm writing these lines, another NEA of about 12 meters (2014 RC) is making a close approach to the Earth just inside the geostationary ring, at a distance from the Earth's surface of 39,900 km.

Nowadays, communication rules the world and everything, even ordinary or insignificant, must be pushed. This is good for the media and the press, as otherwise they might sometimes even have nothing to say, and for the asteroid impact scientists: if they wait for a more spectacular event (e.g. an asteroid impact) to show the taxpayers that they are doing something, they risk waiting too long. It is a pity that they did not have a chance to predict the Chelyabinsk event. It would have been a big promotion for their activities.

Chapter 6

What can we do to best protect the Earth?

To people fidgeting and struggling to save the Earth.
Remember, one way or another Nature will forget about us.

Don't let me be misunderstood. With my criticism of impact monitoring systems, I don't want the reader to think that I consider the Earth safe from the threat of asteroid impact. We know that asteroids have hit the Earth in the past. Impact craters are clearly visible on the Moon and also on the Earth. We have fewer scars than the Moon simply because our planet changes over geological eras (winds, floods, earthquakes and other tectonic movements), while the Moon is dead from this point of view. Moreover, what we currently know of the NEA population tells us that it will surely happen again.

What I am criticizing is the way in which what we know is translated into risk perception. All this recent and compulsive "probability-dropping" by impact monitoring systems is giving a distorted and worrying perception of the risk, more or less as some people did during the Dark Ages with the passage of innocuous comets. Mankind will not die out tomorrow because of an asteroid impact, neither next year, nor in the next thousands of years with an extremely high degree of confidence. We are not under attack, as it may appear with the flourishing of impact threat announcements in the last decade and a half. We know that big and mankind-threatening impacts (kilometer-sized) are extremely, extremely rare (every one million years, on average). Impacts of smaller but locally destructive objects (50 to 200 meters) are more frequent (five hundred to twenty thousand years). Impacts of even smaller objects (meter-sized) are frequent (one to one hundred years), but the chances that they will reach the Earth's surface or even seriously hit a populated area are tiny. Remember that it is estimated that less than 2% of the entire Earth's surface is inhabited by humans.

So, what is the best we can do to protect our lives from the threat of asteroid impact? The answer is: discover, track and catalog as many unknown NEAs as possible. We should spend more time and devote more effort and attention to the discovery of the still unknown NEAs and to reliable orbit computation. This is an ordinary job, and should be done and perceived as such. Astronomers and, above all, impact monitoring people should not behave like Cassandra, warning of unheeded imminent disasters or doom, nor they should consent to be perceived and depicted by the media as Bruce Willis; we are not part of a B-movie[1].

As I stated in chapter 2, governments, national and international space agencies (e.g. NASA) are already on this path. Since the early 90's, many discovery surveys have been financed to discover NEAs. The current goal is to have knowledge of the total population down to $140\,m$ 90% complete, hopefully within the next two decades. That interest has given a tremendous boost to the discovery of NEAs: by the end of the 90's the discovery rate was about 200 per year. Now, more

---

[1]I am referring to the 1998 American science fiction disaster thriller film "Armageddon".

than one thousand objects are discovered every year.

More discoveries obviously mean more objects in the impact monitoring Risk Lists. This is a matter of statistics: with the current rate of discovery (~ 4 NEAs per day, on average) we have a rate of new objects included in the Risk Lists of 3 to 4 per month, on average. If the discovery rate increases, the same happens to the "risky objects" one. It should now be clear to the reader that being included in the Risk Lists has nothing to do with being really and immediately "risky" or "dangerous". It is more akin to belonging to a subset with specific values of orbital parameters: if you double the annual discoveries of NEAs, you will also statistically double the number of NEAs with those specific values of orbital parameters. To discover asteroids is more or less like uniformly sampling colored marbles from a drawer. If you double the extracted marbles, you will surely double the relative number of red marbles. If you go onto the Internet and go through the Risk Lists, you will notice a nice regularity: when you count the number of objects included per year (it is enough to count only objects with a provisional designation starting with the chosen calendar year, e.g. 2013), you will notice that their number is very similar every year (above all, if you consider the last calendar years). Quite a considerable regularity for a system aiming to spot exceptional events.

Unfortunately, the increase of Risk List objects with more discoveries is a nuisance we need to put up with. But now the reader has the tools to understand and critically evaluate what "risky" really means: almost nothing. As I have shown in chapter 4, almost every single object included in the Risk Lists is less "dangerous" than a still unknown NEA of similar size. I do not exclude that with a flood of impact threat announcements in the coming years, no one will care anymore, not even the impact monitoring people themselves. This is called saturation due to useless information.

Regrettably, the discovery and cataloging of a significant percentage of the whole NEA population down to small sizes will realistically require decades, if not a century, of hard observational work and orbit computation. It is a long time to wait and the present-day theoreticians (mainly mathematicians) are getting bored. They want to feel themselves important for society, because nowadays if you are not riding high, then you are nothing. So, they have invented such things as impact monitoring systems, Risk Lists, Torino and Palermo scales.

It could be objected, however, that if an asteroid hits the Earth in the future and the asteroid is already known, then it will be surely and promptly included in the Risk Lists. Yes, this is true. However, as I already said, impact monitoring systems do not tell us right now that the asteroid will hit the Earth in the future with certainty. They cannot really know what nobody knows. At the beginning, when the orbit is poorly known, they will say that the asteroid has a non-zero "impact probability" V_i. This probability will fluctuate with new astrometric data and still during this phase its numerical value means nothing about the actual fate of the asteroid, as I showed previously. The way in which it fluctuates cannot be predicted, otherwise they could know from the beginning whether it will go to 0 or to 1 (the orbital information is the same for everyone, they cannot do magic). Once V_i becomes equal to 1, they can say something meaningful, but at this point every orbit computer (e.g. the Minor Planet Center) already knows that the asteroid will impact the Earth, because "the impact is in the data" and the astrometric data are available to everyone.

Essentially, when nobody can say anything about the actual fate of the asteroid, neither can impact monitoring systems: the V_i value does not add any information. When the impact monitoring systems become aware of the actual fate, every orbit computer has already done so, because the information which lets the monitoring systems see that V_i is eventually equal to 1 is the same that lets every orbit computer see that the asteroid will hit the Earth. By the way, this is precisely what happened with asteroid 2008 TC$_3$. According to the following review of the events posted on The Planetary Society website[2]:

> ...at 06:38 UTC on October 6, astronomers at the University of Arizona discovered an object provisionally called 8TA9D69 that appeared to be on a collision course with Earth. Three other observatories reported sightings within the next few hours - Sabino Canyon in Arizona and Siding Spring Observatory and a Royal Astronomical Society site, both in Australia. Together these four observers provided enough data on the object so that *a Minor Planet Electronic Circular was issued at 14:59 UTC the same day, giving 8TA9D69 the more formal name 2008 TC3, and advising the astronomical community that "The nominal orbit given above has 2008 TC3 coming to within one earth radius around Oct. 7.1 [...]"*. [italics mine]

The impact was actually announced by MPC. We did not need impact monitoring systems to know that the probability was 1. And I am confident that the same would have happened with the Chelyabinsk meteor if only it had been known days, weeks or years before impact.

Nevertheless, I must admit that impact monitoring is not completely useless or misleading. The Risk Lists can be intended as observing priority lists: astronomers are invited to observe these objects in order to improve our knowledge of

---

[2]http://www.planetary.org/blogs/emily-lakdawalla/2008/1684.html

their orbits. No more, no less than this. No words like "dangerous", "risky", or "threatening" should be used to identify these asteroids. There is another useful thing that we can do, although, I believe, it is not as important as NEA discovery, tracking and cataloging: feasibility studies and experiments of asteroid deflection through spacecrafts. This requires a relatively small but well targeted financial investment. I believe that they are worthwhile but, I am sure, we will never have the urgency to build a dedicated mission in the next few centuries or thousands of years, at the least.

Let me conclude this chapter with an anecdote highly representative of the way in which things are run today in almost all the scientific communities. Please focus on the "evocative" side of the story and not on the specific facts (names, etc.).

On the 11th September 2014, the Space Generation Advisory Council's Near-Earth Objects Working Group[3] announced the winner of the seventh annual "Move an Asteroid" technical paper competition: Clemens Rumpf, Germany. In his paper (available at http://arxiv.org/abs/1410.4471) the author describes a software tool under development that allows the calculation of the impact location probability distribution on the surface of the Earth (in the literature, occasionally referred to as a risk or impact corridor) of any asteroid included in the

---

[3]The Space Generation Advisory Council (SGAC) is a non-governmental organization and professional network which "aims to bring the views of students and young space professionals to the United Nations (UN), space industry and other organizations". The Near-Earth Object Project Group (NEO) is dedicated to helping the worldwide planetary defense community observe and track planetary comets and asteroids which have a possibility of colliding with Earth. The group provides a young adult perspective to planetary defense through annual reports, competitions, conference attendance, and public outreach projects related to Near-Earth Objects. (Source: Wikipedia). The "Move an Asteroid" technical paper competition is thought to challenge students and young professionals worldwide to come up with original ideas relating to Near-Earth Objects (NEOs).

impact monitoring Risk Lists. According to the author:

"This work contributes to the risk assessment seg-
ment of the asteroid threat. NASA and ESA maintain freely accessible lists of asteroids that have a
non-zero chance of impacting the Earth in the future. The lists also provide the orbital elements of
these asteroids. However, the potential impact lo-
cations of the asteroids are not available. [...] The
research presented in this paper closes the gap between the information given in the risk lists and
the tools that are able to determine impact effects.
The impact location probability distribution of asteroids in the risk list is determined and a preliminary estimate of casualties is provided highlighting
the global nature of the asteroid threat."

The key result of this study is presented on page 8, Figure 8
(reproduced here in Figure 6.1): it shows the global asteroid
risk map that gives (color coded on a base ten logarithmic
scale) the regions of the globe with higher risk of impact casualties.

Allow me to make some comments. I believe that the results of this paper are misleading and dangerously tenuous
from a political and social point of view. They are based on
very limited statistics: only a subset of the asteroids in the impact monitoring Risk Lists. As I said before, all the asteroids
in the Risk Lists are *absolutely not* the most dangerous ones,
quite the contrary. Almost all the risk comes from the still
unknown NEA population and since the impact location of
potentially impacting asteroids obviously depends upon their
orbits, how can that kind of analysis provide sound, reliable
and useful information when we do not even know how many
and which potentially impacting asteroids are there, let alone
their orbits? Such a highly delicate result as the terrestrial

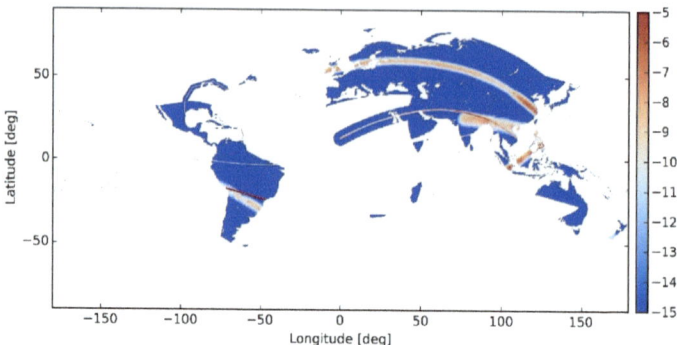

Figure 6.1: From the article's Figure 8 caption "Global asteroid risk map color coded on a base ten logarithmic scale. White
regions are associated with a risk that is effectively zero. The
total risk is evaluated as 29,919 casualties." (Source: ArXiv)

map with high or low risk of impact casualties should not rely
on poor statistics, and least of all on the wrong list of objects.
By the way, in his study the author knowingly includes even
asteroid 2011 AG$_5$, no longer listed in any Risk Lists since the
21st December 2012.

Second. Let me make a cynical and acerbic comment on
the unstoppable sprouting of the prizes, medals and honors being awarded in every scientific community, even the smallest
and most insignificant, especially to young people who have
only just joined the respective clans. These prizes and the
awarding modalities have something between a farcical and an
Orwellian flavor to me. Prizes are like the lure in a greyhound
racing. The community (consciously or unconsciously, I don't
know) uses these tricks to push young moldable minds within
the community to work hard on topics that the community
thinks to be important. Or maybe, all this is only a wider and
more official form of self pep talk, if not self-advertisement. It
is highly probable that the only reason for which most of the

awarded scholars will be remembered (their lifetime achieve-ment) will be just the fact of them receiving the award.

Chapter 7

Brief anatomy of a decade-long rejection

> *But this is an imperfect world, as actors, athletes, priests and presidents soon discover. Writers have known this from the day they read their first review.*
>
> Thomas Fleming

The core of this book, my criticism of the monitoring systems' "impact probability", has already been published[1] as a seven-page paper in the peer-reviewed journal *Chance* in 2013. This journal is an official magazine of the American Statistical Association. In fact, the basic idea (a discomfort, I would say) about the main result of impact monitoring systems dates back to 2002, when a first version of the paper saw the light. You can still read the forefather of the *Chance* article at http://arxiv.org/abs/physics/0212058: if the reader goes through it, he/she will see that, apart from a less thorough analysis of the then existing data and some "cosmetic" imperfections, it already contains all the key points of my criticism and basically the same results/conclusions.

---

[1]http://www.tandfonline.com/doi/full/10.1080/09332480.2013.794610 #.VAbOebuXXUI. The paper is also freely available at http://arxiv.org/abs/1010.5125.

I guess that the reader is asking why wait so long to publish a paper? Well, the story of my paper is a really troubled one. I submitted the paper (in all its versions) to five different scientific journals in the time span of 11 years. Four of them rejected the paper and one (obviously, the last one) eventually accepted it for publication. The story of my paper is also a story of stubbornness. In this chapter I want to piece together all the facts, the temporary defeats and the final "victory" that shaped this story. And I will do so also by quoting passages from the referees' reports on my submitted paper. I must tell the reader that to profitably read this chapter he/she should have already read chapter 4.

This chapter is not intended to be an auto-celebration. In it, I want to show the reader with a practical case, experienced in the first person, how rubbery the rubber wall of peer-review can become when you go against the mainstream, even a small and sectoral one like that of asteroid impact monitoring science, even with a plain and blatantly true argument. It must be said that most of the time reviewers are in good faith, and I do not want to summon up conspiracy theories. It is just the way in which such an impersonal thing as the mainstream affects our ability to discern: you end up unconsciously filtering out what is at odds with it. In principle, I have no prejudice against peer-review. On paper it is a powerful tool for authors to have an independent check of the validity of what they say and to improve their work (formally and substantially). However, what happens in reality is that there are a lot of researchers around, even in small research fields, and a lot of papers to be refereed (nowadays there is a lot more research than in the past, and current researchers publish more papers per year than 50 or 60 years ago). Thus, it is not unusual that a researcher is asked by journal editors to review several papers per year and, sometimes, even on topics that he/she has not mastered at all. This is not always a manageable load, to the point that it is not possible to give the right share of time to the work under review. Moreover, and this is worse, when researchers are chosen to be referees, they often feel compelled to act like a severe and pedantic schoolmarm who marks with a red pen everything that she does not immediately understand and, above all, every linguistic inaccuracy as if it were the Plague[2].

With the "professionalization" and "industrialization" of Science in the last century, people working in it need a sort of "validation" or "certification" of what they say and do, also to be hired and/or climb the academic ranks, and this task is fulfilled by peer-review. But let me be very clear: in the knowledge of Nature, agreement by peers means nothing. Nature does not need the agreement of reviewers to behave as it behaves. Every true scientist knows very well that he/she has to go toe to toe directly with Nature, not with what other human beings say.

---

[2]I said that I have no prejudice against peer-review, but sometimes reviewers do have against authors. Recently, the distinguished American planetologist A. W. Harris (of whom I will tell later in this chapter) kindly asked me to be the second author of a paper reviewing our joint 14 years activity in the estimation of the NEA size distribution with a probabilistic approach we devised in 2001. As a matter of fact, I was the first author of the 2001 paper, but then he improved the methodology with a clever debiasing technique and kept on updating the estimate with new data. The 2015 paper was submitted to *Icarus* and eventually accepted for publication. In the reviewing phase, a reviewer commented: *"Parts of the manuscript have awkward constructions in terms of the language. Most notably the very first sentence in the abstract in passive with the second sentence in active. The language of the manuscript would improve if a native English-speaker (such as the lead author) would read the entire manuscript and correct the most obvious issues."* That comment suggests that the reviewer thought that the "awkward" part of the manuscript was written by the non-native English-speaker author (i.e. me). The truth is that the entire paper was written by the American planetologist. I am pretty sure that had my name been John Smith or James Fisher, that kind of remark would not have been made.

Although peer-review is the rule today among scientific journals (pay-per-publish journals are a matter I don't want to deal with here), I may safely say that the quality of published papers has inexorably declined in the last 60 years[3]: 99.9% of the tens of thousands of papers published every year are rubbish and will be soon forgotten, even by the authors themselves (I have no problem admitting that probably even my papers will be within this 99.9%).

Here, I do not want to hide my responsibility for what happened with my paper. Above all with the first version of 2002, I admit that the probably not-so-linear phrasing and the English could have been at the origin of some misunderstandings. However, later versions of the paper were almost the same as the published one. So, why for some reviewers is a paper difficult to understand because it is poorly written, while for others the very same paper deserves publication after minor changes? The way in which it was written could have played a role in it being rejected, but a minor one. What baffles me is that a significant fraction of the reviewers (above all those working in the astronomy, planetology and astrophysics world) simply did not understand my plainly true argument.

October, 2002. *Astronomy & Astrophysics* (EDP Sciences)

I began to feel uncomfortable with the theory behind asteroid impact monitoring science very early on. I knew that there was something wrong with their "impact probability" and with the interpretation the Clan gave to it. Less than three years after the official birth of the first impact monitoring system (1999, CLOMON, at the University of Pisa), I wrote a nine-page

---

[3]In this regard, I strongly recommend the watching of a brief talk by Gerald Pollack entitled "The Ills of Science". https://www.youtube.com/watch?v=Tz0bC_4_xLo

paper in which all the key points against the monitoring systems' "impact probability" can already be found. The paper was entitled "A posteriori reading of Virtual Impactors impact probability" and it is still available at the arXiv repository (see the link above).

I was extremely confident in the validity of my thesis and decided to submit the manuscript to the international journal *Astronomy & Astrophysics* (A&A). I edited the text according to the rules of the journal and in October 2002 I sent an electronic copy to the editorial office of A&A.

On the 10th December 2002, I received an email from the former Editor-in-Chief, Claude Bertout, communicating that the paper had been sent to a competent referee (not anonymous) who recommended rejection on the grounds that its scientific content was not sufficient to warrant publication in A&A. The competent referee was Prof. Andrea Milani. I also received his report. I quote it in full below:

Title: A posteriori reading of virtual impactors impact probability

Authors: G. d'Abramo

I have read this paper, and I do not believe it contains anything correct. The main problem is that the author appears not to understand the very definition of probability. The idea, on which he insists in Section 2, that a value of the impact probability as computed by CLOMON and/or Sentry cannot be considered reliable because it changes with time is completely wrong. The probability of a future event, which is in fact the outcome of a deterministic dynamic, is simply a measure of our ignorance on the initial conditions of the orbit. Thus the probability MUST change every time the information available on the asteroid in question changes.

If we try to compute a probability for an event, which is independent from observational information and therefore is time-independent, it can only be some a priori value, which is deduced from independent information. E.g., if we do not look in the sky at all for Near Earth Asteroids, the probability of an impact is just the so-called background probability, which is deduced from the counts of lunar (and also terrestrial) impact craters, not from any observation of asteroids/comets. If the author refuses to use the observation information, which is contained in the computations of impact probability, this is the only conclusion he would be able to draw: that to detect an asteroid impacting next year has a low probability, the same as the yearly rate of impact averaged over the last hundreds of million years. Besides, this is a tautology: if there is no new information, because we refuse to use the available one, the probability does not change.

The main problem, of course, is how to USE the observation information, to detect an impacting asteroid, if any. If we refuse to do this, why are we looking? Indeed, whatever the long term probability, we do not know if the next imapct [*sic*] of a given size is going to be next year rather than in 10,000 years. The purpose of the whole exercise (astrometry, orbit computation, impact monitoring) is to see if our times are at variance from the average, that is if we are going to be hit within our lifetime. If we are happy with the idea that this is unlikely, we can just stop.

The computation of the impact probabilite probabilites [*sic*] is a difficult subject, using many delicate assumptions on the statistical properties of the observational errors, and quite a few mathematical shortcuts; thus it is possible, actually desirable, that some other qualified author contributes to the theory and criticizes how these computations are done. However, the author of this paper does not even mention the computational technique and the underlying assumtions [*sic*] of the probability computations. Besides, the author has not used even a single number of impact probability, computed eiter [*sic*] by CLOMON or by Sentry, in his argument: these value do not appear in his formulas. Probably the author does not even have an archive containing a record of these data. How can he state that the result of some computation is wrong without knowing the result itself?

In conclusion, I strongly recommend that this paper is not published.

I was at the beginning of my (now non-existent) career in academic research and it was a hard blow. I was and have always been sure of my point, but then I felt frustrated: how can a full professor in Mathematics, a leading researcher in Celestial Mechanics like him, not understand such a plain and obvious result? A week later I decided to reply directly to Prof. Milani, trying to explain my point and clear up all misunderstandings. I tried to explain that my criticism was not of the way their probability is calculated but of its meaning and interpretation. I hinted at arguments not very much different from those I have used through this book. But this was difficult for me for two reasons: first, I was talking to a guru in his field of research and, second, all that I said in my paper still seemed plainly true to me. It is difficult to defend your position when it is blatantly true to you. If the report of Prof. Milani seems convincing to the reader, I invite him/her to re-read the last two chapters of this book.

By the way, when he says: *"The purpose of the whole exercise (astrometry, orbit computation, impact monitoring) is to see if our times are at variance from the average, that is if we are going to be hit within our lifetime"*, I would simply reply that the probability that he is successful with his exercise is given again by the background impact probability. No more, no less. At any rate, I suddenly realized that I would never have been able to publish my paper in any of the astrophysics journals in the world. Thus, I buried it in a drawer, and there it remained for 8 long years.

October, 2010. *Studies in History and Philosophy of Modern Physics* (Elsevier)

To use an image, there was still a fire smoldering under the ashes. I continued to be extremely confident in the validity of my arguments. And the arrogance and scarce sensibility of the rising Clan pushed me not to give up. Meanwhile, my research interest was ripening toward thermodynamics, probability, history and epistemology of physics. I got into the habit of reading *Studies in History and Philosophy of Modern Physics* (SHPMP) and I thought that this journal could give greater attention to subjects related to the definition and interpretation of an impact probability.

On the 8th October 2010, I submitted the paper (slightly updated and closer to the version eventually published in *Chance*) to SHPMP. It was entitled "Earth impact threat of newly discovered asteroids: an *a posteriori* analysis".

I had to wait several months. Finally, on the 1st April 2011 (April Fool?) the Editors replied that they had received two reports on my submission: of these, one was more positive than the other. The Editors decided to give more weight to the critical report, in view of the strict standards of SH-

PMP (another whole book could be written about this way of doing things, somewhere between psychology and epistemology). They were not able to publish the paper as it was, and required substantial revision. Actually, this was a substantial improvement compared with the brush-off given 8 years before by A&A.

Reviewer #2 (anonymous) concluded his/her report with these words:

> In conclusion, let me re-emphasize that this paper certainly deserves rapid publication in SHPMP after minor improvements have been made.

Since, this time, the Editors gave me a chance to reply and see my paper published, I gave a step by step response to the more critical reviewer, Reviewer #1. I must tell the reader a curious story that happened before receiving the Editor's email, through which I became aware of the identity of Reviewer #1. This reviewer sent his review report directly to me, more than one month before the Editorial Office communication. He was Clark R. Chapman, a famous planetologist working at the Southwest Research Institute, Boulder CO. Below, I quote just his general comments:

> I am not familiar with this journal, so it is difficult for me to assess whether this manuscript is typical of articles published in the journal. However, I would expect not. The article is written in an unnecessarily technical way and would be understood, I expect, only by a small number of specialists in asteroid observations and celestial mechanics, none of whom are likely to be readers of this journal. I believe that the topic **could** be written in a way that would be readable and potentially

interesting to people who I imagine to be readers of this journal. It involves an inherently interesting topic. But there is another problem: I think the author's conclusions are obviously wrong. It is possible, instead, that I have misunderstood the argument, but I think it is more likely that I have understood it well enough to say that it is likely to be fallacious. I have written specific comments below that should help the author understand why I think it is wrong (or where I have misunderstood his argument, which would point to how to fix the writing).

Thus, I would recommend rejection of the article. But the author should be given the chance to demonstrate that I have misunderstood his argument, if that is the case. In which case, he should explain how, and then perhaps he could completely rewrite this article in a way that would be much more widely accessible to readers who are not experts in asteroids.

I replied to every specific comment, but I prefer not to quote anything else here because in order to understand my reply you would need to find and accurately read the submitted paper: this is an endeavor that I would not wish even on my worst enemy. It is enough to say that I spent all the time clearing up all the misunderstandings and explaining again my plainly true point in other words. After having sent (full of hope) my detailed reply (9 pages long, about the length of the paper itself) to the Editorial Office, I had to wait for another six months to receive the final decision. Which was baffling. It even seemed for a moment that I had been able to persuade Reviewer #1. But it was not enough. Let me quote the report in full:

Title: Earth impact threat of newly discovered asteroids: an a posteriori analysis, Studies in History and Philosophy of Modern Physics

Dear germano,

The Editors have extensively discussed this paper. Finally, following the comments on the revised version of the attached report, they have decided that the paper is not historically and philosophically significant enough to warrant publication in SHPMP. Publication in a more specialized journal seems appropriate.

Sincerely,

G.S.

Managing Editor

Reviewer #1: The author has responded to my review of his original submission by often agreeing with my criticism and making appropriate changes to the manuscript. In particular, the manuscript is now more accessible and less obscure. Also, technical issues with the first manuscript, which I took to be errors, the author generally believes were not errors but were written in misleading ways. In the revision, these parts of the manuscript have either been rewritten in more understandable ways or omitted from the revised manuscript. I consider the English to present some minor issues that impede clear understanding (e.g. the frequent phrase in the manuscript "new more" I first read as "no more", but the author actually means "additional new").

The editor has assured me that the rather technical nature of this article is not a problem for readers

of this journal (a journal I am not familiar with),
so I drop that objection.

The editor has asked me to concentrate on the issue of whether the author makes a clear conceptual
and philosophically interesting point. I have difficulty answering this question. The discussion, in
the first part of the manuscript, of the probabilities of impacts by Near Earth Asteroids may be
novel and interesting to readers of this journal...but
this isn't new. What is new is the derivation of
the probability W. *I think the derivation is correct* [emphasis mine]. But, as I understand W, I
don't find it to be especially interesting or useful.
Of course, I am different from readers of SHPMP,
since this topic is one in which I am expert, and
what interests me and what interests people not
knowledgeable about this field may differ.

I would have thought that the author would motivate interest in the utility of W by using it in
some concrete examples in order to show how it
provides a different perspective from the VI impact probability. But he doesn't. In the original
version he applied it to the case of Apophis, which
I found to be very confusing; in his response to my
review, the author claims that the application was
correct, but he has chosen to omit discussion of
Apophis, so that example no longer appears in the
manuscript. In the revised manuscript, the author
appears to say (near the top of pg. 6) that he will
discuss the case of 1999 AN_{10} in Sect. 2, but he
then fails to discuss it there or anywhere later in
the paper. My suggestion is that the author provide some cogent examples where consideration of
W provides a different perspective on a particular

case of a possible impact than mere consideration of the impact probability.

As for whether derivation of W would be deemed to be a philosophically interesting point by readers of SHPMP, I simply cannot say, since I have little idea about what might interest readers of the journal. Therefore, I make "No Recommendation" on this form. Should the reviewer form not accept "No Recommendation", I will then choose "Minor Revision" (referring to my idea that the author motivate interest in W by providing some examples), but I really have to leave it to the editor, who is familiar with the journal, its readers, and what interests them.

Just when I felt the concrete possibility of having my paper published, I had to start again, nearly from scratch.

September, 2011. *Monthly Notices of the Royal Astronomical Society* (Wiley)

After the second rejection, I did not lose heart. I immediately found another Journal and submitted the very same paper, with all the improvements I had made in accordance with the two previous reviews. I chose the British journal *Monthly Notices of The Royal Astronomical Society* (MNRAS). At the end of September 2011, I sent the manuscript to MNRAS. On 1st November 2011, the Editorial Board replied with a rebuff. These were the Editor's comments:

Editor's comments:

I sent the manuscript for review to an expert referee with considerable expertise in statistical analysis and a strong interest in the terrestrial impact record. The attached report outlines a number of fundamental difficulties with the definition and use of the quantity W, which constitutes the main focus of the paper. The referee's main objection, which I see as being a valid and fatal flaw in the paper's central argument, is that the NEA population from which most of the Virtual Impactor population arises, is almost certainly not the population responsible for the cratering record.

The report also highlights a number of flaws in the statistical analysis, including a number of rather fundamental conceptual errors such as the confusion between impact rates and probabilities following equation (1), and the erroneous description of the ratio W itself as being a probability. As the referee notes, any information that contributes usefully to our knowledge of the impact probability in the Bayesian sense is most profitably used to improve V_i itself.

I hope that the advice given in the attached report may prove useful if you decide to develop the idea further in future. For the moment, however, I concur with the referee's view that the manuscript does not establish the usefulness of the quantity W with sufficient rigour to justify further consideration for publication in MNRAS in anything like its present form.

I also recommend that any future submission to a scientific journal should avoid the rather inappropriate and sensationalist style of language used in the manuscript, as noted by the referee. This did not influence the decision to reject, but I certainly would have asked for it to be changed if further revision was invited.

In particular, the remark *"The referee's main objection, which I see as being a valid and fatal flaw in the paper's central argument, is that the NEA population from which most of the Virtual Impactor population arises, is almost certainly not the population responsible for the cratering record"* was puzzling to me. I still have difficulty understanding what on Earth it has to do with my paper. Actually, I still have difficulty in understanding what it means in general. Taken literally, it is unintentionally silly: it is plainly true that the NEA population from which most of the Virtual Impactor population arises is NOT the population responsible for the cratering record, since the latter is gone, vaporized with the impact.

A posteriori, I must admit that the Editor was right in one thing: the confusion between impact rates and probabilities following the equation of the background impact probability. This confusion was actually clarified in the published version in *Chance*. However, this confusion, contrary to what the Editor (and the Reviewer) said, did not affect at all the definition and the validity of probability W. It must be said, to be fair, that if the whole community of planetologists usually call "background impact probability" what is actually an impact rate, the confusion is not only mine. At any rate, it is clearly a confusion in the words used, not in the intended meaning.

November, 2011. *Astrophysics & Space Science* (Springer)

After the quick and burning exchange with the Editorial Office of the last journal, I did not even have the time to feel discouraged and soon I decided to resubmit the same paper to yet another journal. I had grown used to rebuttals to the point that anger overcame frustration. This time I chose the *Astrophysics & Space Science* (A&SS) journal. On the 29th November 2011, I submitted the paper to A&SS. On the 16th

January 2012, I received an email from the Editorial Office
communicating that my paper, based on the advice received,
was rejected. Attached to the email, there were two review
reports.

Reviewer #1, who did not require anonymity, was very
kind. Although he was critical of the paper, he gave me the
opportunity to clarify all the issues obscure to him (more or
less like Reviewer #1 of SHPMP). He even offered to provide
me with a more up-to-date plot of his own. This is what he
wrote in a passage of his report:

> But I reiterate, I am not sure I understand all that
> is being presented, so I invite the author to clar-
> ify what I may have missed, especially so that a
> less specialized reader might better appreciate the
> points of the paper.

It is exactly this that I meant when I said that peer review
could be an important tool for authors. But you have to be
lucky to meet genuine scientists as reviewers. By the way, Reviewer #1 was Alan W. Harris, a Senior Researcher, formerly
at JPL.

Reviewer #2 was less lenient. His/her arguments were not
different from those given by previous critical referees. But let
me anyway quote some passages from his/her report:

> ...
>
> There are many problems with the English lan-
> guage that would need to be corrected but in light
> of the following general comments, I did not see
> the point of listing them.
>
> ...
>
> More than half of the paper provides a high level
> background on asteroids and impact assessments

but the explanations are not comprehensive enough to be useful. The technical details were omitted by the author since he notes that the discussion would be too technical for the intended audience.

...

The a posteriori impact probability "W" does not seem useful to this reviewer. The probability that the impact probability will go to 1 is already fully represented in the standard impact probability that is currently in use. The "W" parameter really does nothing more than note that the smaller asteroids impact more frequently – a point that is already obvious since there are so many more small ones. Evaluating the probability of the impact probability going to 100% without reference to the current impact probability is not helpful.

...

For a paper that I consider inappropriate for publication, I would normally provide suggestions for making the article acceptable. However, I frankly do not see how this paper could be made acceptable for publication.

I hope that the reader, at this point of the book, is able to see for him/herself that Reviewer #2 utterly missed the point about W when he/she said *"The "W" parameter really does nothing more than note that the smaller asteroids impact more frequently – a point that is already obvious since there are so many more small ones."*

Besides, does the reader remember when I said that most referees act like a severe and pedantic schoolmarm marking with a red pen everything that she does not immediately understand and, above all, every linguistic inaccuracy? Exactly.

Since the Editors rejected the paper squarely, I did not reply, but I'd have liked to. Above all to Reviewer #2, who made a lot of specious claims. Who reviews the reviewers?

April, 2012. *Chance* (American Statistical Association, Taylor & Francis)

Before submitting my paper again for the fifth time, I admit that I had a moment of weakness. I thought for a while that I'd put my paper away in a drawer again. This time forever. In that period, I used to define what had passed through the crossfire of more than seven referees, my "best-refereed unpublished paper". However, only a few weeks later, I made the decision to submit the paper again (with the title "How to Defuse Earth Impact Threat Announcements") to a journal which did not have anything to do with astronomy and astrophysics academia: *Chance* journal. After all, the only time I had even come close to publication was with a history and philosophy of physics journal.

This choice was a fortunate one. First, because my paper was eventually published. Second, because I had the chance to talk to very qualified referees (this time, the Editor chose 4 referees!), who understood and eventually agreed with what I was saying. I felt that I was not alone. They were also extremely supportive and collaborative.

Chapter 8

A humble invitation to humbleness

Mankind is doomed to boredom. Everything we do is urged by it: power-seeking, will to do and to know. And whatever we get, we call it progress.

Given the actual timescale of significant impact events here on the Earth, there is a lot time for impact monitoring scientists to get bored and, accordingly, there is plenty of time for them to make progress.

I hope to have given within this book a picture of the asteroid impact issue that is complete enough to allow the reader to form his/her own idea: should one worry about asteroid impacts? Should one be on the alert for every impact threat announcement? Should one care and somehow make a contribution to asteroid impact science? Or should one ignore all this stuff and worry about problems closer to home and even more threatening? I leave the answer to the reader. I want him/her to know that I am open-minded about that.

I can understand people[1] who find sense in their own life

---

[1] Like professional string theorists, particle and theoretical physicists, impact monitoring scientists, mathematicians, etc.

only when they devote themselves to a rather exotic area of knowledge, detached from everyday life. If it happens also that this employment requires a lot of training, sacrifice and the use of technical jargon that is even more detached from everyday life, then these people and the rest of humanity usually end up thinking that what they are doing is of the highest level and extremely important for the world. They are perceived and perceive themselves as divines with the privilege of having been initiated into an obscure and almost sacred discipline.

But, who can say what is more or less important for humanity? Take a look at your own life. Are you sure you can say with certainty what has been crucial in your life and what has not, among all the things that have happened to you? Are you sure that if you were able to delete events from your past that appear insignificant to you, your life would be the same now, or better?

What I want to say, before becoming too mystical, is that we should strive to think critically about everything that is currently mainstream. I would humbly remind the reader that not too long ago people grew rich and powerful, affecting for a while the fate of mankind, simply because they were strong supporters of silly theories like that of a geocentric universe[2], the mainstream of the period. Outsiders had trouble, so to speak. And those who think that nowadays we have eventually and definitely reached the light, do not know what science really is: an always tentative knowledge. And they do not even know who man really is: an always limited being.

---

[2]Although mainly unrelated to the topic of this book, let me make this brief excursion into epistemology: to anyone who speaks of the "unreasonable effectiveness of mathematics", I usually reply that even the mathematics of the untrue geocentric theory must be acknowledged as a facet of the "unreasonable effectiveness of mathematics". Mathematics is not physics; I believe deeply that mathematics is able to provide a convincing formal framework for every theory we are willing to believe; mathematics is not the Queen, but the Servant of the Sciences.

Obviously, I am not against Science or Knowledge, but I have grown annoyed hearing impact monitoring scientists crying out loud and taking for granted that what they do is urgent and absolutely necessary for the present and the future of mankind. And if one soundly criticizes their work, they react as if you are offending directly the Goddess of Science, more or less as people who criticized the earthly business of priests were treated as heretics during the Middle Ages. Let them do their job, OK. But I wish to remind these people that just as the greatest discoveries are almost always unexpected, the greatest dangers always come from what we do not yet know. So, do what you think is important, but always be down-to-earth!

Chapter 9

Ebb and flow

Somewhere in the Middle Ages...

If you avow publicly that you do not believe in God and you are a woman, you are surely a witch. If you are a man, you are a heretic. At any rate, you will end up burning at the stake.

Somewhere today...

If you act coldly towards or are uninterested in the alleged wonders of current "Science" and Technology or even publicly criticize the multi-millionaire budget of some experiments (LHC, Space missions, etc), you are an ignorant, obscurantist troglodyte who poses a severe threat to the progress of humanity.

It is certainly better to live today, but this is a meager consolation.

Selected Bibliography

- Einstein, Albert. Out of My Later Years: The Scientist, Philosopher, and Man Portrayed Through His Own Words. Philosophical Library (1950).

- The Spaceguard Survey: Report of the NASA International Near-Earth-Object Detection Workshop. Morrison, D., Editor, 1992, NASA. https://archive.org/details/nasa_techdoc_19920025001

- López-Corredoira, Martin. The Twilight of the Scientific Age. BrownWalker Press (2013).

- Unzicker, Alexander and Sheilla Jones. Bankrupting Physics: How Today's Top Scientists are Gambling Away Their Credibility. Palgrave Macmillan Trade (2013).

- Chapman, C.R. and D. Morrison, 1994. Impacts on the Earth by asteroids and comets: assessing the hazard. *Nature* 367: 33.

- Chapman, C.R. 1999. The asteroid/comet impact hazard. Case study for Workshop on Prediction in the Earth Sciences: Use and Misuse in Policy Making. July 10-12, 1997, National Center for Atmospheric Research, Boulder, CO, and September 10-12, 1998, Estes Park, CO. www.boulder.swri.edu/clark/ncar799.html.

- Milani, A., Chesley, S.R., Chodas, P.W., Valsecchi, G.B. Asteroid Close Approaches: Analysis and Potential Impact Detection. In Asteroids III, 2003, W. F. Bottke Jr., A. Cellino, P. Paolicchi, and R. P. Binzel (eds), University of Arizona Press, Tucson.

- Harris, A.W. 2008. What Spaceguard did. *Nature* 453: 1178.

- D'Abramo, Germano. 2013. How to Defuse Earth Impact Threat Announcements. *Chance* 26(2), 17-23.

- Morrison, D., Harris, A.W. Sommer, G. Chapman, C.R. and Carusi, A. Dealing with the impact hazard. In Asteroids III, 2003, W. F. Bottke Jr., A. Cellino, P. Paolicchi, and R. P. Binzel (eds), University of Arizona Press, Tucson.

www.ingramcontent.com/pod-product-compliance
Lightning Source LLC
Chambersburg PA
CBHW040905180526
45159CB00010BA/2928